In Situ Hybridization

Principles and Practice

236956/2709UR
AB   HBI

Edited by

JULIA M. POLAK and JAMES O'D. McGEE

Oxford   New York   Tokyo
OXFORD UNIVERSITY PRESS

Oxford University Press, Walton Street, Oxford OX2 6DP
Oxford New York Toronto
Delhi Bombay Calcutta Madras Karachi
Kuala Lumpur Singapore Hong Kong Tokyo
Nairobi Dar es Salaam Cape Town
Melbourne Auckland Madrid
and associated companies in
Berlin Ibadan

Oxford is a trade mark of Oxford University Press

Published in the United States
by Oxford University Press Inc., New York

First published 1990
Reprinted 1991, 1992

British Library Cataloguing in Publication Data
In situ hybridization: principles and practice.
1. Medicine. Neurochemistry
I. Polak, Julia M. (Julia Margaret) II. McGee, J. O'D.
III. Series
612'.8042
ISBN 0-19-261906-3

Library of Congress Cataloging in Publication Data
In situ hybridization: principles and practice/edited by Julia M. Polak and J. O'D. McGee.
1. Nucleic acid hybridization. 2. Molecular biology—Technique.
I. McGee, James O'Donnell. II. Title.
QH506.036 1990
574.87'328—dc20 89-28454 CIP
ISBN 0-19-261906-3

Printed by Information Press, Oxford

Preface

'If there be fuel prepared, it is hard to tell whence the spark shall come that shall set it on fire'. Francis Bacon, '*Of seditions and troubles*'. Essays (1625).

Pathology is undergoing a revolution, with the arrival of new imaging methods for revealing the basic mechanisms of disease. Data provided by immunocytochemistry and electron microscopy are now being complemented by molecular biology. One of these techniques, *in situ* hybridization, visualizes in intact cells the genes and mRNA involved in protein synthesis. There were many difficulties to overcome in working out the methodology, but the technique is now established in many research, and some clinical laboratories. It has been rapidly adopted by pathologists and biologists.

This book forms part of a series on the practical aspects of technology at the forefront of pathology: it deals with *in situ* hybridization and its applications to the study of mechanisms underlying cellular function and disease. The authors who have contributed to this book are recognized experts and their chapters cover all aspects of *in situ* hybridization from the basic principles of molecular biology through the various methodological advances to the most up-to-date applications in developmental biology, virology, cytogenetics, and pathobiology. There is currently more emphasis on DNA than on mRNA techniques and their applications. This imbalance will be corrected with improved methodology for investigations on mRNAs in health and disease.

We are grateful to Oxford University Press for asking us to put together in a single publication these expositions of the state of the art, and latest advances, in this intriguing and fascinating laboratory technique.

'Great cultural changes begin in affectation and end in routine'. Jacques Barzan, *The house of intellect* (1959).

London and Oxford
September 1989

J. M. P.
J. O'D. McG.

Contents

Contents

Contributors

BUPENDRA BHATT

University of Oxford, Nuffield Department of Pathology and Bacteriology, John Radcliffe Hospital, Oxford OX3 9DU, UK.

SAVILE BRADBURY

University of Oxford, Department of Human Anatomy, South Parks Road, Oxford OX1 3QX, UK.

MICHAEL A. W. BRADY

Amersham International plc., Corporate Research, Pollards Wood Laboratories, Nightingales Lane, Chalfont St. Giles, Bucks HP8 4SP, UK.

VICTOR T-W. CHAN

Oncology Division, Department of Medicine, Beth Israel Hospital, Harvard Medical School, 330 Brookline Avenue, Boston MA 02215, USA.

ANTHONY P. DAVENPORT

Clinical Pharmacology Unit, Department of Medicine, University of Cambridge, Level 2, F & G Block, Addenbrooke's Hospital, Hills Road, Cambridge CB2 2QQ, UK.

MARTIN F. FINLAN

Amersham International plc., Corporate Research, Pollards Wood Laboratories, Nightingales Lane, Chalfont St. Giles, Bucks HP8 4SP, UK.

DAVID M. J. FLANNERY

University of Oxford, Nuffield Department of Pathology and Bacteriology, John Radcliffe Hospital, Oxford OX3 9DU, UK.

SALLY J. GIBSON

Department of Histochemistry, Royal Postgraduate Medical School, Hammersmith Hospital, Du Cane Road, London W12 0NN, UK.

C. SIMON HERRINGTON

University of Oxford, Nuffield Department of Pathology and Bacteriology, John Radcliffe Hospital, Oxford OX3 9DU, UK.

HEINZ HÖFLER

Institute of Pathology, Technical University of Munich, School of Medicine, Ismaningerstrasse 22, D-8000, Munich 80, FRG.

A. H. N. HOPMAN

Department of Pathology, University Hospital Nijmegen, Geert Grooteplein, Zuid 24, 6525 GA Nijmegen, The Netherlands.

RICHARD LATHE

LGME-CNRS and U184-Inserm, 11 Rue Humann, 67085 Strasburg, Cedex, France.

JAMES O'D. McGEE

University of Oxford, Nuffield Department of Pathology and Bacteriology, John Radcliffe Hospital, Oxford OX3 9DU, UK.

DEREK J. NUNEZ

Clinical Pharmacology Unit, Department of Medicine, University of Cambridge, Level 2, F & G Block, Addenbrooke's Hospital, Hills Road, Cambridge CB2 2QQ, UK.

JULIA M. POLAK

Department of Histochemistry, Royal Postgraduate Medical School, Hammersmith Hospital, Du Cane Road, London W12 0NN, UK.

F. C. S. RAMAEKERS

Department of Pathology, University Hospital Nijmegen, Geert Grooteplein, Zuid 24, 6525 GA Nijmegen, The Netherlands.

C. G. TEO

Department of Virology, Royal Postgraduate Medical School, Hammersmith Hospital, Du Cane Road, London W12 0NN, UK.

G. P. VOOIJS

Department of Pathology, University Hospital Nijmegen, Geert Grooteplein, Zuid 24, 6525 GA Nijmegen, The Netherlands.

DAVID G. WILKINSON

Laboratory of Developmental Biochemistry, National Institute for Medical Research, The Ridgeway, Mill Hill, London NW7 1AA, UK

1

Basic background of molecular biology

V. T-W. CHAN, C. S. HERRINGTON, AND J. O'D. McGEE

Molecular Cloning

Molecular cloning is a technique to propagate a DNA segment with 'absolute' sequence purity in large quantity. In a typical experiment of gene cloning, a prerequisite is preparation of the DNA sequence to be propagated. Usually this sequence is in a complex population of DNA molecules that are either cDNAs (complementary DNA) synthesized from messenger RNAs (mRNA) or genomic DNAs. For cDNA cloning, total RNA is first extracted, and then mRNA is purified from this RNA preparation. Since most eukaryotic mRNAs occur naturally in a polyadenylated form, they can be purified by oligo (dT) cellulose. At high salt concentration, the poly(A) tails of mRNAs re-anneal to cellulose-linked oligo (dT) chains and are immobilized. After other RNA species are washed off, mRNA can be recovered by elution with low-ionic-strength buffer. The sequence of interest can be further enriched by several methods such as size fractionation. cDNA copies of mRNAs are synthesized by reverse transcriptase. The immediate product of the reaction is an RNA–DNA hybrid. The RNA strand can be destroyed by chemical or enzymatic reactions, to which DNA is resistant, leaving a single-stranded cDNA that can be converted to a double-stranded form by a second DNA polymerase reaction. This reaction depends on the formation, by the reverse transcriptase reaction, of a transient self-pairing structure in which a hairpin loop at the 3'-terminus is stabilized by enough base-pairing to allow initiation of synthesis of the second strand. The hairpin and other single-stranded portion of the cDNA molecule can be trimmed by treatment with a single-strand specific enzyme, S1 nuclease, giving rise to a fully duplex molecule.

For the cloning of genomic sequences, DNA is first extracted and randomly digested or mechanically sheared to a desired size. If a particular gene is to be cloned, partial purification or enrichment before cloning can reduce the number of recombinants that have to be screened. Unfractionated DNA fragments representing the entire genome can also be used.

Sequences of interest are introduced into cloning vehicles known as vectors prior to their propagation. These vectors possess genetic functions that are utilized in the selective propagation and subsequent selection of recombinants (i.e. vector-insert combinations).

There are four classes of cloning vehicle now in common use: plasmids, bacteriophages, phagemids, and cosmids. More recently, very large fragments of DNA have been inserted into yeast artificial chromosomes (YAKs). Plasmids are replicons that are stably inherited in an extrachromosomal state. They are widely distributed throughout prokaryotes. They are double-stranded, closed circular DNA molecules varying in size from 1 kilobase pair (kb) to greater than 200 kb. Often, plasmids contain genes coding for enzymes that, under certain circumstances, are advantageous to the bacterial host. Among the phenotypes conferred by different plasmids on bacteria are resistance to antibiotics, degradation of complex organic compounds, production of antibiotics, enterotoxins, restriction enzymes, and modification enzymes. Most plasmids are derived from the plasmid pBR322 (Fig. 1.1(a)). This was originally synthesized by combining the naturally occurring plasmids pSC101, ColE1 and RSF124 (Old and Primrose 1985). pBR322 has been modified by the introduction and deletion of various genetic functions which increase their versatility as cloning vehicles. Among these functions are the *lacZ* gene, which allows the selection of recombinants by colour (e.g. the pUC series, Fig. 1.1(b)); RNA promoter sites, such as SP6 (Fig. 1.1(c)) and T3/T7 (Fig. 1.1(d)); and multiple cloning sites, which permit the acceptance of exogenous DNA cut with a wide variety of restriction enzymes.

The second class of vector consists of the bacteriophages. These are viruses that have prokaryotes as their host. The two most widely used bacteriophages in molecular biology are bacteriophage λ and the filamentous phage M13.

Bacteriophage λ is a virus of *E. coli*. It is genetically complex but has been extensively studied and, indeed, was developed as a cloning vector early in the history of gene manipulation. The DNA of bacteriophage λ, in the form in which it is isolated from the phage particle, is a linear duplex molecule of 48.5 kb, the entire DNA sequence of which has been determined. At each end there is a short single-stranded 5′ projection of 12 nucleotides: these are complementary in sequence. These cohesive termini can associate to form the *cos* site by which the phage DNA adopts a circular structure when it is injected into its host cells. Genes of the central region are concerned with the recombination and the process of lysogenization in which the circularized chromosome is inserted into its host chromosome itself. Most of the central region, including these genes, is not essential for phage growth and can be deleted or replaced without seriously impairing the vegetative cycle. This dispensability is crucially important in the construction of vector derivatives of the phage. Derivatives of wild-type

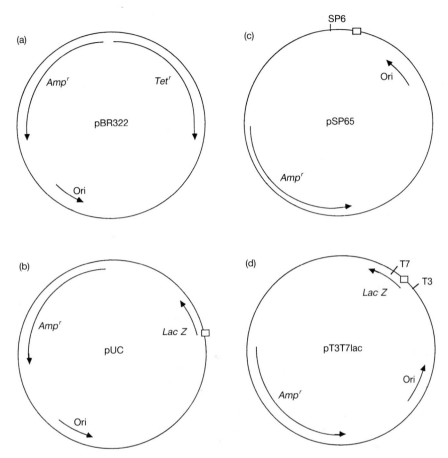

Fig. 1.1. (a) The plasmid pBR322 contains ampicillin and tetracycline resistance genes. Insertion of exogenous DNA into one of several unique restriction sites in either gene allows selection of recombinant molecules. (b) The pUC series was derived by excision of the *tet'* gene and insertion of the *lacZ* gene, which codes for the enzyme β-galactosidase. A multiple cloning site (MCS) was inserted into this gene without interrupting the open reading frame and therefore selection of recombinants can be performed on the basis of loss of enzyme activity (X-gal selection), making the colonies colourless rather than blue in the presence of substrate. (c) The pSP6 series was derived by insertion of the Salmonella phage SP6 RNA polymerase promoter upstream of the MCS of pUC12. This insertion removed most of the *lacZ* gene and therefore selection by X-gal is not possible. (d) pT3T7lac: the logical extension of the *pSP6* series was to insert T3 and T7 RNA polymerase promoters on either side of the MCS. These are arranged in opposite orientation, thus allowing the production of sense and anti-sense RNA transcripts from the same recombinant. The insertion of these promoters does not interrupt the *lacZ* gene and therefore recombinants can be selected using X-gal (see legend of (b)).

Ori = origin of replication;

☐ = multiple cloning site;

Amp' = ampicillin resistance gene;

Tet' = tetracycline resistance gene;

LacZ = β-galactosidase gene.

phage have either a single target site at which foreign DNA can be inserted, or have a pair of sites defining a fragment that can be removed and replaced by foreign DNA. Since phage λ can only accommodate about 5 per cent more than its normal complement of DNA, vector derivatives have been constructed with deletions to increase the insert space within the genome. The shortest λ DNA molecules that produce plaques of nearly normal size are 25 per cent deleted. Apparently, if too much non-essential DNA is deleted from the genome, it cannot be packaged in phage particles efficiently. This can be turned to advantage, for if the replaceable fragment of a replacement type vector is removed by physical separation, or is effectively cleaved by another restriction enzyme, then the deleted vector can give rise to plaques only if a new DNA segment is inserted into it. This accounts for positive selection for recombinant phages.

Increasing use is being made of derivatives of the filamentous bacteriophage M13 as vectors. These combine the advantages of the vegetative phage replication cycle with the versatility of the *lacZ* gene function and multiple cloning sites. The single-stranded form in which M13 exists in the phage particle can be used to facilitate DNA sequencing by the dideoxynucleotide method of Sanger. More recently, vector systems known as phagemids, which behave as plasmids but have the functions of M13, which allow the isolation of a single-stranded form for DNA sequencing, have been formulated. They also possess RNA promoter sites (T3/T7) for the generation of RNA transcripts.

Plasmids have been constructed that contain a fragment of λ DNA including the *cos* site. These vectors have been termed cosmids and can be used as cloning vectors in conjunction with *in vitro* packaging systems because only a small region in proximity to the *cos* site is required for recognition by the packaging system. Packaging the cosmid recombinants into phage coats imposes a desirable selection upon their size. With a cosmid size of approximately 7 kb, the insertion of 30–45 kb of foreign DNA is possible—much more than a phage λ vector can accommodate. After packaging *in vitro*, the particles are used to infect a suitable host. The recombinant cosmid DNA is injected and circularized like phage DNA, but replicates as a normal plasmid without the expression of any phage functions. Transformed cells are selected on the basis of a vector drug-resistance marker. Used in this way, cosmids provide an efficient means of cloning large pieces of foreign DNA.

In molecular cloning, there must be a way to cut the vectors and the DNA fragments to be cloned at desirable sites and then join them together to produce recombinant molecules. Restriction endonucleases (type II) recognize particular target sequences in duplex DNA molecules and break the polynucleotide strands within, or close to these sequences. In fact, the DNA fragments to be cloned are usually prepared by either partial or complete digestion by these enzymes.

After cutting the DNA molecules (both the insert and vehicle DNA), the DNA fragments are joined by the reaction of T4 DNA ligase (from T4 phage-infected *E. coli*) to create recombinant molecules. Having prepared a vector–insert combination containing the sequence of interest, the bacterial cell must accept the recombinant and allow it to replicate. Treatment of *E. coli* with $CaCl_2$ allows cells to take up DNA from bacteriophage λ. $CaCl_2$-treated *E. coli* cells are also effective recipients for plasmids. Almost any strain of *E. coli* can be transformed with plasmids, albeit with varying efficiency. Many bacteria contain restriction systems that can influence the efficiency of transformation. Although the complete function of these restriction systems is not yet clear, one role they do play is the recognition and degradation of foreign DNA. For this reason, it is usual to use a restriction minus mutant of *E. coli* as the recipient cell. In general, large DNA recombinants transform *E. coli* less efficiently, on a molar basis, than small DNA molecules. The approach that can be used to circumvent the problem of low transformation efficiency is to package recombinant DNA into phage particles *in vitro*.

Placing the recombinant DNA in a phage coat, when either bacteriophage λ or cosmid is used as cloning vector, allows it to be introduced into the host bacteria by the normal process of phage infection, i.e. phage adsorption followed by DNA injection. λ phage DNA in concatemeric form, produced by a rolling circle replication mechanism, is the substrate for the packaging system. The principle of packaging *in vitro* is to supply the ligated recombinant DNA with high concentrations of phage head precursors, packaging proteins and phage tails. Practically, this is most efficiently performed in a very concentrated mixed lysate of two induced lysogens, one of which is blocked at the prehead stage while the other is prevented from forming any head structure. In the mixed lysate, genetic complementation occurs and exogenous DNA is packaged. Covalently joined recombinant DNAs can also be prepared in concatemeric form in the ligation reaction and thus be packaged *in vitro*.

Identification of the clones carrying the desired recombinant is the most important step in a molecular cloning experiment. All vector molecules carry suitable genetic markers or properties. Plasmids and cosmids carry drug resistance or mutation markers and in the case of phage vectors, plaque formation is itself the selection marker. Genetic selection for the presence of vectors is a prerequisite stage in obtaining the recombinant population. This can be refined to distinguish recombinant molecules and non-recombinant parental vectors. Insertional inactivation of a drug-resistance marker, or of a gene such as β-galactosidase for which there is a colour test (see above and Fig. 1(b)), are examples of this. With certain replacement type λ vectors, and with cosmid vectors, size selection by the phage packaging system selects for recombinants.

Selection of clones can be based on the information contained within the desired sequence. If an inserted sequence in the desired recombinant is expressed, the genetic selection may provide the simplest method for isolating clones containing the gene. For example, cloned *E. coli* DNA fragments carrying a biosynthetic gene can be identified by complementation of non-revertible auxotrophic mutations in the host *E. coli* strain.

Immunochemical detection has been used successfully to identify the clones in which the inserted gene is expressed and not degraded by the host. A particular advantage of this method is that genes that do not confer any selectable property on the host can be detected, but it does, of course, require that specific antibody is available. This method mainly depends on three considerations: (1) an immune serum containing several IgG types that bind to different antigenic determinants on the antigen molecules, i.e. it is polyclonal; (2) antibodies absorb very strongly to plastics such as polyvinyl, from which they are not removed by subsequent washing; and (3) IgG antibodies can readily be labelled with reporter molecules. Transformed bacteria are plated on agar in a Petri dish. A replica is prepared because the colonies are killed in the subsequent procedures. The bacterial colonies are then lysed and antigens from the positive clones are released. A sheet of polyvinyl precoated with appropriate antibody (unlabelled) is applied to the surface of the plate, whereupon the antigens complex with the bound antibodies. The labelled IgG can then react with the bound antigens via antigenic determinants at sites other than those involved in the initial binding of antigens to the immobilized antibodies. Positive colonies can be identified by the detection of the labelled (e.g. alkaline phosphatase) IgG. The desired clones can then be recovered from the replica plate.

A recombinant detection method employing hybridization with DNA isolated and purified from the transformed bacteria has also been developed. However, this has been completely superseded by the method of Grunstein and Hogness (1975), who have developed a screening procedure to detect DNA sequences in transformed colonies by replica filter hybridization with radioactive DNA probe. This procedure can rapidly determine which colony among thousands contains the required sequence. Modification of the method allows screening of colonies plated at very high density.

The colonies to be screened are first replica plated onto a nitrocellulose filter disc that has been placed on the surface of an agar plate prior to inoculation. A reference set of these colonies on the master plate is retained. The filter bearing the colonies is removed and treated with alkali so that the bacterial colonies are lysed and their DNA is denatured. The filter is then treated with proteinase K to remove proteins and leave denatured DNA bound on the nitrocellulose in the form of a DNA-print of the colonies. DNA can be fixed firmly by baking the filter at 80 °C. The defining, labelled DNA is hybridized to the DNA-print and the positive colonies are identified by autoradiography. The positive clones can be recovered from the reference plates. Modification of this procedure can be applied to phage plaques by placing a nitrocellulose filter on the surface of agar plates, making direct contact between plaques and filter. The plaques contain considerable amounts of unpackaged phage DNA which binds to the filter and can be denatured, fixed, hybridized, and detected.

This method has the advantage that several identical DNA prints can be easily made from a single agar plate; it thus allows the screening to be performed in duplicate, and hence with increased reliability, and also allows a single set of recombinants to be screened with two or more probes. The great advantage of the hybridization method is its generality. It does not require expression of the inserted sequences and can be applied to any sequence provided a suitable probe is available. As originally described, the method employed RNA probes. However, with slight modification, DNA probes can be used. These screening procedures are powerful and can be applied generally. Using these procedures, it is now possible to isolate any gene sequence from virtually any organism easily, if a suitable probe is available.

Nucleic acid hybridization

In a hybridization reaction, a molecular probe has to be prepared. After a desired genomic clone has been identified, it is usually subcloned into plasmids to remove the repetitive sequences from the original clone. In this way, the coding sequence is cloned into another vector and thus the specificity of the hybridization signal is increased. For example, in the original cloning step, the gene interested is cloned into either phage λ or cosmid. After identifying the desired clone, a fragment containing the coding sequence (or part of it) is subcloned into a plasmid. The plasmid subclone can then be used as a hybridization probe.

Preparation of molecular probes

Large quantities of plasmid can be prepared using simple methods. The yield varies from less than one milligram to several milligrams per litre of bacterial culture, depending on the nature of the construct. In general, plasmids under relaxed control give higher yields whilst the yields of those under stringent control are much less. Bacteria can either be grown to saturation or their growth can be stopped at late log phase by the addition of chloramphenicol to amplify the copy number of the plasmid (for those under relaxed control). The yields of these two methods do not differ significantly. However, the former procedure is much simpler than the latter because there is no need to monitor the growth of bacteria. The cell walls of bacteria are first cleaved with lysozyme followed by the lysis of their cell membranes. The bacteria can be lysed either chemically or physically. Chromosomal DNA and cellular proteins are removed from the bacterial lysate by high-speed centrifugation, leaving plasmids, RNA, and trace amount of cellular proteins in the supernatant. Plasmids can then be purified by density-gradient ultracentrifugation.

RNA has a higher density than DNA which in turn has a higher density than protein. Caesium chloride is usually used to form the gradient. After overnight spinning, a density gradient is formed along the length of the centrifuge tube. For the purification of supercoiled plasmid, ethidium bromide (EtBr) has to be added to the caesium chloride solution of bacterial lysate. Ethidium bromide binds by intercalating between DNA base pairs and in so doing causes the DNA to unwind. Supercoiled plasmid has no free ends and can only unwind to a limited extent, thus limiting the amount of EtBr bound. Linear DNA fragment or nicked plasmid have no such topological constraints and can therefore bind more of the EtBr molecules. Because the density of DNA/EtBr complex decreases as more EtBR is bound, and because more EtBr can be bound to a linear DNA or a nicked plasmid than a supercoiled plasmid, the supercoiled plasmid has a higher density at saturating concentration of EtBr. Therefore, supercoiled plasmid can be separated from other macromolecules present in the bacterial lysate by caesium chloride gradient ultracentrifugation in the presence of EtBr.

Several methods can be used to prepare plasmid DNA. Of these methods, alkaline lysis is relatively simple and efficient. Lysozyme is used to break down the cell walls of bacteria and the cell membrane is then lysed by sodium dodecyl sulphate (SDS) and sodium hydroxide. Acidic solution of potassium acetate is added to neutralize the alkali. At the same time, chromosomal DNA and most of the cellular proteins are precipitated by the high concentration of potassium acetate. Nucleic acids are precipitated from the clear lysate with isopropanol and the supercoiled plasmid is purified by equilibrium density centrifugation. This method can be used to prepare plasmids of varying size as well as cosmids. The final product is of high purity.

Alternatively, there is a variety of methods for plasmid purification which avoid caesium chloride centrifugation (Maniatis *et al.* 1989). These employ proteinase K and phenol/chloroform extraction, and RNase digestion, to remove protein and RNA respectively. Although DNA produced by these methods is adequate for most purposes, it is inevitably of reduced purity compared with that isolated by caesium chloride centrifugation because of remaining protein and bacterial chromosomal DNA.

Hybridization reaction

Solution/Filter hybridization

Molecular hybridization is a reaction in which single-stranded target sequences in solution, on filters, or within tissues, and a complementary molecular probe anneal to form double-stranded hybrid molecules. In the case of DNA hybridization, both target sequences and molecular probe are originally double-stranded. Before hybridization, they must be denatured, to render them single-stranded. This can be achieved by alkali or heat. In the case of nucleotide analogues in which the reporter molecule (e.g. biotin) is attached to the base by an ester linkage, heat denaturation must be used, because alkali will remove the reporter molecule by alkaline hydrolysis. After denaturation, the mixture of the resultant DNA fragments is incubated under conditions that favour re-annealing of single-stranded fragments. For RNA hybridization, denaturation of target sequences is not theoretically necessary but is advantageous because RNA molecules assume a secondary structure in solution. Denaturation, therefore, increases the proportion of fragments accessible for hybridization. In single-phase, i.e. solution hybridization, the optimal temperature for renaturation is approximately 25 °C lower than the melting temperature (T_m) of the hybrid molecules. The T_m of a duplex DNA molecule, which is defined as the temperature at which 50 per cent of duplexes are dissociated, depends on the following variables: the proportion of guanidine and cytidine nucleotides (%G–C), the length of the duplex in base pairs (L), the concentration of monovalent cation (M) and the amount of formamide (F) in the reaction mixture. These variables are linked by an equation derived by experiment for DNA association in solution (see, e.g. Hames and Higgins 1985).

$$T_m = 81.5 + 16.6 \log M + 0.41(\%G + C) - 0.72F - 650/L$$

An additional correction is applicable to mismatched hybrids (see below). The stringency of hybridization and washing determine the degree to which mismatched hybrids are permitted to form. The T_m of mismatched hybrids (T'_m), i.e. those formed between heterologous sequences, is lower than that for matched sequences. This reduction occurs in a predictable fashion according to the equation:

$$T'_m = T_m - \chi(\%\text{mismatch}), \qquad \text{where } \chi = 0.5 \rightarrow 1.4 \,°\text{C}$$

The precise relationship depends on the nucleotide sequence of the hybrid as mismatch of GC-rich segments reduces the T_m by a greater amount than mismatch of AT-rich regions, i.e. $\chi = 0.5$ for poly(AT) and 1.4 for poly(GC). For most sequences, the value of χ lies between these two extremes: e.g. 0.95 ± 0.05 °C for human papillomaviruses. Thus, if hybridization is carried out at a temperature between T'_m and T_m, only matched sequences will hybridize. However, the rate of re-annealing is maximum at $T_m - 25$ °C and falls as the temperature of hybridization approaches T_m. A compromise is therefore required and conventional 'stringent' conditions (hybridization in 50% formamide, 2 × standard saline citrate (SSC) at 37 °C) represent $T_m - 17$ °C, allowing rapid re-annealing but only distinguishing between sequences sharing less than ~ 83 per cent of their sequences (for $\chi = 1$). Higher stringency is then achieved by post-hybridization washes to values closer to the T_m according to the equation above.

The use of oligonucleotide probes is described in detail elsewhere (see Chapter 5) and the following discussion of stringency conditions employed in non-isotopic *in situ* hybridization (NISH) refers to the use of nick translated probes on paraffin sections (see Chapter 12).

In situ hybridization

The hybridization and washing conditions employed in *in situ* hybridization (ISH) have been formulated by assuming that nucleic acids within cells and tissues behave in the same way as those in solution. However, cellular material contains many other limiting factors, such as cytoskeletal structures, that impair diffusion of probe to target. Aldehyde fixation increases these differences both by further impeding diffusion and by cross-linking nucleic acids to each other and to associated proteins. Archival formalin-fixed biopsy material is increasingly being investigated by NISH and any deviation of the behaviour of nucleic acids from those in solution is of importance. It has been noted that denaturation of nucleic acid within aldehyde-fixed material (cells or tissue) has to be performed at 90–95 °C in the presence of 50% formamide and $2 \times SSC$ despite the fact that the T_m derived from the equation above is 58 °C (assuming $G + C = 50$ per cent). Thus, the T_m of matched target DNA duplexes is elevated.

The ability of ISH to distinguish between closely related sequences has also been assumed to be determined by the equations derived from solution kinetics (see above). This assumption has been investigated recently by analysing the infection of archival biopsies with human papillomaviruses (HPV). These provide an ideal system in which to study probes for closely homologous sequences. In particular, HPV types 6 and 11 share an average of 82 per cent of their sequences. Although no case of double infection with these two viruses has been reported by filter hybridization analysis of 123 positive cases, several studies have indicated a high degree of positivity with both probes in the same lesion by ISH. This discrepancy has been shown to be due to cross-hybridization of HPV6 and 11 probes with target sequences in paraffin sections (Herrington *et al.* 1990). The T_m of matched and mismatched hybrids has been estimated in paraffin sections by increasing the stringency of post-hybridization washes. As the stringency of post-hybridization washing is determined solely by the T_m of the hybrids, this gives a true measure of T_m. Using this approach, it was found that the linear relationships noted between salt, formamide and T_m in solution are also valid in archival biopsies (Herrington *et al.* 1990). It is, however, not possible to measure the T_m, as defined above for DNA association in solution, by NISH. Firstly, the end point of 50 per cent dissociation is not measurable and, secondly, short probes have to be used in NISH to allow access of probe to target. These considerations have prompted the introduction of the term T'_m to denote the melting temperature derived by NISH analysis using nick translated probes. The end point used to define this term was defined as the midpoint of the temperatures at which signal was either absent or just retained under particular washing conditions. This parameter is, nevertheless, likely to approximate closely the classical solution T_m as the melting profile of the longer probe segments remaining near the point at which all signal disappears is narrow. Equations relating T'_m to salt and formamide concentration have been constructed:

$$T'_m = 85.7 + 6.4 \log M - 0.42 \,(\%\text{formamide})$$

The constant 85.7, takes account of GC content (41 per cent for HPV6 and 11). Thus, lowering salt and raising formamide concentration is a less effective method of reducing the T_m than in solution. This is likely to be due to the stabilization of probe–target hybrids by factors other than those being measured, namely cellular constituents. The linear relationship with salt holds over the range $0.1–2 \times \text{SSC}$ and for formamide from 10 to 60 per cent. Extrapolation outside these ranges and to other systems may not be valid (see below).

In conclusion, the stringency of hybridization and washing conditions is determined by many properties of the biochemical environment of nucleic acid hybrids. The equation derived by analysis of nucleic acid interaction in solution does not hold for nucleic acids in tissues, particularly those which have been aldehyde cross-linked. However, the manipulation of salt and formamide concentration does have a predictable effect on the behaviour of DNA–DNA hybrids formed between closely homologous human papillomaviruses within tissues. This deviation is likely to be due to the effect of other cellular constituents on the stability of hybrids, and therefore the optimal stringency for a given system should be determined by experiment. It should be stressed that all equations quoted above have been derived by experiment, either in solution or by NISH on clinical biopsies. Only viral DNA–DNA hybrids have been discussed but RNA–DNA hybrids (e.g. the detection of mRNA using DNA probes) and RNA–RNA hybrids (e.g. the detection of mRNA using RNA probes) have also been studied by NISH. Although limited information is available on these hybrids in solution (Hames and Higgins 1985), there has been no formal investigation by NISH. Similarly, non-viral DNA may not behave in an identical manner to the HPV sequences discussed above. Preliminary experiments with human genomic DNA–DNA hybrids indicate the generality of the T'_m equation (Herrington and McGee, unpublished).

Application of molecular biology to the study of human disease

Genetic manipulation techniques can be applied to the study of many different human diseases such as microbial and viral infection, genetic diseases, and neoplastic disorders. In viral infection, a typical example is the study of the molecular biology of HIV, which is responsible for AIDS. The viral genome has been cloned and the functions of different segments of the genome were identified in a relatively short time. The difference between HIV and other retroviruses is that its genome contains at least six open reading frames, and therefore potentially six genes, while other retroviruses usually only encode for three or four genes. Furthermore, there are two transacting genes which regulate the replication of the virus. This technique has also been used in studies of other viruses such as hepatitis B virus, Epstein–Barr virus, hepatitis delta virus, etc. (see Chapter 9). Understanding regulatory mechanisms of the replication of these infection agents is an important step in combating these diseases.

Genetic diseases that are caused by lesions in certain genes are obvious candidates for analysis using recombinant DNA technology. The most extensively studied genetic diseases are globin-related lesions such as sickle-cell anaemia and thalassaemias. In sickle cell anaemia, it is well known that the cause is an A-to-T transversion at codon 6 of the β-globin gene, resulting in a substitution of valine for glutamic acid in the protein product. With the use of a restriction enzyme, Mst II, this mutation can be detected by a technique called restriction fragment length polymorphism (RFLP) because the enzyme cleaves the normal allele whilst the mutated allele is resistant to this cleavage. This technique can be applied to prenatal diagnosis of sickle-cell anaemia in high-risk populations.

Thalassaemias are a group of disorders in which there is an imbalance in the synthesis of globin chains owing to low or totally absent synthesis of one of them. Many α-thalassaemias are caused by gene deletions, although several non-deletion forms have been identified. However, β-thalassaemias are complex and more than 20 different lesions have been identified. In some cases, the structural changes of the gene are caused by mutation of only one or two bases in the DNA. It is possible to detect a number of these base changes directly in the DNA if the mutations lie within the recognition sites of restriction enzymes. It is also possible to detect not only the inheritance of the gene but rather the inheritance of a DNA fragment that is closely linked to the gene. Thus, if a non-functional gene is associated with a novel restriction site (RFLP), and the fetus inherits the restriction site, it will also inherit the non-functional gene in the absence of recombination. Two different sorts of polymorphism have been defined. In one case, the polymorphism is found associated with the same sort of mutated gene in many people, i.e. allelic-specific linked polymorphism. In the other case, however, different polymorphisms may be found associated with the mutated genes in different people, i.e. linked polymorphism. These polymorphisms may provide a means for diagnosis.

In addition, with the techniques of genetic manipulation, the genes responsible for other genetic diseases such as cystic fibrosis, Huntington's disease, and Duchenne muscular dystrophy have been identified or may be identified in the near future. The possibility of performing gene therapy is also under investigation. However, the major problem of gene therapy is the stable establishment and consistent expression of the introduced genes. Its success would obviously be useful for the treatment of genetic disorders.

In cancer research, recombinant DNA technology is also widely used. Several approaches are adopted to investigate the nature of cancers and the pathway of carcinogenesis. Of the different approaches, activation of cellular oncogenes has been studied extensively. From the results of these studies, a hypothesis of multi-step carcinogenesis involving oncogene activation has been described (for details see Chan and McGee 1987).

The gene responsible for the childhood tumour retinoblastoma has been identified (Friend *et al.* 1986) and that responsible for Wilm's tumour will probably be cloned in the near future. It has been proposed that these tumours are caused by at least two genetic alterations at the respective gene loci. The first one is an inheritance of a recessive mutation at the gene locus, resulting in a predisposition to the tumours. The second one is a somatic event leading to the loss of the normal allele, or then, reduplication of the mutated allele (development of hemizygosity or homozygosity). The retinoblastoma gene is located on chromosome 13q14 while the gene involved in Wilm's tumour is probably located on chromosome 11p13 (Orkin *et al.* 1984).

Treatment of cancer is another subject in cancer research. Multi-drug resistance of certain cancers soon after chemotherapy is a major problem of cancer treatment. This resistance is not only to the selective drug but also to other seemingly unrelated agents. Over-expression or amplification of a membrane glycoprotein, mdr (also called p-glycoprotein), may be found coincident with multi-drug resistance of tumour cells. During selection for increased levels of resistance, expression of its mRNA is increased simultaneously with amplification of this gene. Over-expression of the *mdr* gene from a drug-sensitive cell line is sufficient to confer multi-drug resistance on cell lines that are originally drug-sensitive, suggesting mutation of the gene is not required for this resistance (Gros *et al.* 1986).

References

Chan, V. T.-W. and McGee, J. O'D. (1987). Cellular oncogenes in neoplasia. *J. Clin. Pathol.*, **40**(9), 1055–63.
Friend, S. H., Bernards, R., Rogelj, S., Weinberg, R. A., Rapaport, J. M., Albert, D.

M., and Dryja, T. P. (1986). A human DNA segment with properties of the gene that predisposes to retinoblastoma and osteosarcoma. *Nature*, **323**, 643–6.

Gros, P., Neriah, Y. B., Croop, J. M., and Houseman, D. E. (1986). Isolation and expression of a complementary DNA that confers multidrug resistance. *Nature*, **323**, 728–31.

Grunstein, M. and Hogness, D. S. (1975). Colony hybridization: a method for the isolation of cloned DNAs that contain a specific gene. *Proc. Natl Acad. Sci. USA.*, **72**, 3691–5.

Hames, B. D. and Higgins, S. J. (eds) (1985). *Nucleic acids hybridisation: a practical approach*. IRL Press, Oxford, Washington DC.

Herrington, C. S., Graham, A. K., Burns, J., and McGee, J. O'D. (1990). Discriminative stringency conditions for detection of human papillomaviruses (HPV) in clinical biopsies by non-isotopic in situ hybridisation using biotin and digoxigenin labelled probes. *Histochem. J.* (In press.)

Maniatis, T., Fritsch, E. F., and Sambrook, J. (1989). *Molecular-cloning—a laboratory manual*. Cold Spring Harbor, New York.

Old, R. W. and Primrose, S. B. (1985). *Principles of gene manipulation*. Blackwell Scientific, Oxford.

Orkin, S. H., Goldman, D. S., and Sallan, S. E. (1984). Development of homozygosity for chromosome 11p markers in Wilm's tumour. *Nature*, **309**, 172–6.

Note added in proof

cDNA libraries (from a few cells) are now being constructed by the polymerase chain reaction (PCR). Clones as long as 1.5 Kb have been isolated in this way.

Reference

Belyavsky, A., Vinogradova, T., and Rajlewsky, K. (1989). PCR-based cDNA library construction: general cDNA libraries at the level of a few cells. *Nucl. Acids Res.* **17**, 2919–34.

2

Principles of *in situ* hybridization

HEINZ HÖFLER

Introduction

Over the last three decades techniques employing immunological and molecular biological techniques have had increasing impact in pathology and biology. The introduction of immunohistochemistry and Western blotting as research and diagnostic tools opened a new era of pathobiology and enabled a better understanding of cell and tumour biology. Over the past decade, the knowledge of molecular biology has expanded dramatically with the development of recombinant DNA techniques and sensitive methods for detection of specific DNA or RNA sequences by molecular hybridization. Generally, hybridization can be performed in solid supports, in solution (*in vitro*), and on tissue sections or cell preparations (*in situ*). The hybridization techniques on solid supports involve the immobilization of extracted and/or electrophoretically separated DNA fragments onto a filter (nitrocellulose or nylon membranes) followed by hybridization to a specific DNA probe (Southern blotting). Similarly, different classes of RNA isolated from a particular cell population or tissue can be identified after electrophoretic separation and transfer onto a solid support prior to hybridization (Northern blotting). If specific DNA or RNA sequences are to be quantified regardless of their length, dot (slot) blot analysis enables the rapid processing of up to one hundred or even more samples on a single membrane. Hybridizations to DNA or RNA can also be performed directly on agarose gel, or in microfuge tubes. These techniques do not require the steps of DNA (RNA) transfer onto filters, and will, therefore, be used more frequently. Although molecular hybridization *in vitro* or to membrane-bound nucleic acids isolated from a particular cell population or tissue allows the identification of different classes of DNA and RNA, it tells us little about the distribution of specific sequences in individual cells. *In situ* hybridization (ISH)—also referred to as hybridization histochemistry or cytological hybridization—is a technique that, in contrast to the other methods, enables the morphological demonstration of specific DNA or RNA sequences in individual cells in tissue sections, single-cells, or chromosome preparations. Hence, ISH is the only method that allows one to study the cellular location of DNA and RNA sequences in a heterogeneous cell population.

15

In situ hybridization was introduced in 1969 (Buongiorno-Nardelli and Amaldi 1969; Gall and Pardue 1969; John *et al.* 1969) and has been used primarily for the localization of DNA sequences. In more recent years ISH has been applied to the localization of viral DNA sequences, mRNA and chromosomal gene mapping. Today, ISH is a powerful research tool in the hands of biologists and pathologists. In many laboratories, ISH is already a routine procedure, particularly in the diagnosis of viral diseases. With the steadily growing number of cloned nucleotide sequences and the number of potentially useful probes, the fields of application for ISH are growing rapidly. Because of its high specificity, ISH will become increasingly important in several areas of biomedical research including developmental biology, cell biology, genetics and pathology (Coghlan *et al.* 1985; Höfler 1987, Matthews and Kricka 1988).

Theoretical background of *in situ* hybridization

ISH is based on the fact that labelled single-stranded fragments of DNA or RNA containing complementary sequences (probes) are hybridized to cellular DNA or RNA under appropriate conditions forming stable hybrids. Since ISH represents the morphological application of primarily molecular biological techniques (for details see Chapter 1), protocols for ISH were originally developed by laboratories in the field of virology and endo-crinology, in which recombinant DNA technology had already had a considerable impact on research activities. Optimal conditions for hybridi-zation to purified DNA or RNA extracts on solid supports have been studied extensively. However, with the use of tissue sections or cell preparations, additional problems such as non-specific binding resulting in high background and reduced signal-to-noise ratio were encountered.

The sensitivity of ISH depends on the following variables: (1) the effect of tissue preparation on retention and accessibility of cellular (target) DNA or RNA; (2) type of probe construct, efficiency of probe labelling and sen-sitivity of the method used for signal detection; and (3) the effect of hybrid-ization conditions on the efficiency of hybridization (see Chapter 1).

Fixation and preparation of tissue

Optimal fixation and tissue preparation should retain the maximal level of cellular target DNA or RNA while maintaining optimal morphological details and allowing sufficient accessibility of probe. In contrast to the rather stable DNA, mRNA is steadily synthesized and degraded enzymati-cally. Consequentially, tissue prepared for RNA localization should be fixed or frozen as soon as possible after surgical excision, and the time between excision and adequate fixation has to be taken into account in each

case when the results of ISH are interpreted (see below). For the localization of DNA, the type and concentration of fixative is not of major
influence. On the other hand, for RNA localization, the type, time and
concentration of the fixative are significant, if loss of RNA is to be
minimized. Paraformaldehyde, a cross-linking fixative, was successfully
used in several studies (Brigati *et al.* 1983; Hafen *et al.* 1983; MacAllister
and Rock 1985; Höfler *et al.* 1986). Unlike other fixatives (e.g. glutaraldehyde), paraformaldehyde does not cross-link proteins so extensively as to
prevent penetration of probes. Precipitating fixatives such as acetic acid–
alcohol mixtures or Bouin's fixation are favoured by other groups (Gall and
Pardue 1971; Manuelidis *et al.* 1982; Berge-Lefranc *et al.* 1983). For
mRNA localization, we prefer fixation of tissue in buffered 4% paraformaldehyde for 1–2 h, followed by immersion in sucrose prior to
freezing. Tissue is then kept in liquid nitrogen until cutting in a cryostat or
vibratome. Alternately, tissue can be snap-frozen immediately after excision, stored in liquid nitrogen, and cut. Sections are briefly (10 min) fixed in
4% paraformaldehyde, air dried and stored at −70 °C. Although morphology is not as well maintained, the latter procedure has the advantage
that the tissue can be used for ISH and DNA and/or RNA extractions.
Tissue sections and single-cell preparations must adhere well to specially
treated glass slides to avoid loss of tissue during the hybridization procedure. Gelatin chrome alum coating of glass slides (Gall and Pardue
1971), or treatment with poly-L-lysine (Huang *et al.* 1983) or silanized
slides (Burns *et al.* 1987) are the most widely used procedures. Formalin
fixation and paraffin embedding is the routine procedure in pathology
laboratories for biopsy preparation; it allows the detection of DNA and
mRNA of higher copy number in most instances. The general reduction of
hybridization efficiency in paraffin sections may be due to reduced accessibility of probe to target, resulting from increased cross-linking of protein in
paraffin sections compared with frozen sections (Campell and Habener
1987), loss of mRNA during the embedding procedure, or more likely to
lack of unmasking of DNA/RNA by proteolysis (see Chapter 12).

Pretreatment of sections with a detergent and/or proteinase digestion is a
standard procedure in almost all published protocols to increase probe
penetration and accessibility, particularly with paraffin sections. Triton
X-100 and RNAse-free proteinase K are commonly used, although Brigati
et al. (1983) recommend pronase rather than proteinase K. Lawrence and
Singer (1985) extensively studied the rate of RNA retention, influence of
prolonged exposure to different fixatives and proteinase digestion on
single-cell preparations. They concluded that paraformaldehyde fixation
gave the highest specific signal and that detergent and protease pretreatment is not necessary. This may be so for myoblasts in culture, but
does not apply to paraffin sections (Herrington *et al.* 1990). Some workers
immerse the sections in acetic anhydride prior to hybridization to reduce

background (Hayashi *et al.* 1978). According to most authors, and from our own experience, this step does not improve signal-to-noise ratio significantly.

The description of protocols for fixation and pretreatment of cells for gene localization on chromosomes is beyond the scope of this contribution. For details and further references, see Chapters 10 and 11, Harper and Saunders (1981), Pardue (1986), and Hopman *et al.* (1988).

Probes

Basically, labelled DNA and RNA probes can be employed to localize DNA and mRNA (Table 2.1). There is agreement that the optimal length of probes for tissue penetration and high hybridization efficiency is 50–300 bases; this is a rough approximation which has not been scientifically evaluated (McGee, unpublished). For special applications such as localization of genes on chromosomes or when 'networking' for signal enhancement (Lawrence and Singer 1985) is required, probes of 1.5 kb should be applied.

Table 2.1. Probes for *in situ* hybridization

DNA	RNA
dsDNA	sscRNA
Synthetic oligonucleotides	
ssDNA	

1. *Double-stranded DNA probes,* usually labelled by nick-translation or random priming are most often used. Nick-translation employs the enzymes DNase I and DNA polymerase I and can be performed with the specific insert only, or with whole plasmid containing the specific insert (Maniatis *et al.* 1982). Comprehensive studies to find the optimal probe concentration, length of nick-translated probe fragments, and time course of hybridization, have been undertaken for cells in culture (Lawrence and Singer 1985). A second way to label double-stranded DNA probes is random primed synthesis of DNA with random primers utilizing Klenow fragment of DNA polymerase (Feinberg and Vogelstein 1983). In contrast to nick-translation, it is necessary to cut the double-stranded DNA template prior to labelling. This method allows the generation of probes with high specific activity but the yield of labelled probe is restricted to approximately 50 ng for each reaction. Therefore, random priming is not recommended for ISH studies in which large numbers of tissue sections are used, because approximately 10 ng of probe is required per each slide.

2. *Synthetic oligodeoxyribonucleotides* can be prepared conveniently by DNA synthesizers and have several advantages over cloned DNA probes:

(a) a consistent and higher specific activity; (b) the possibility of synthesizing probes from amino-acid sequences when the total DNA sequence is unknown; and (c) the ability to generate discriminating sequences for similar genes. An additional advantage is the option of constructing different probes of a particular sequence to prove hybridization specificity. Labelling of oligonucleotides is usually performed by *end labelling techniques*, either by 5'-end labelling with T4 polynucleotide kinase, or 3'-end labelling with terminal deoxynucleotidyl transferase. The main disadvantage of this technique is the low specific activity of probes. For further details, see Chapter 5.

3. *Single-stranded cDNA probes* became possible with the introduction of the cloning vector M13 (Varndell *et al.* 1984). The main reason for the restricted practical use of M13-derived ssDNA probes is probably difficult vector construction and probe synthesis. Recently, with the introduction of the polymerase chain reaction (PCR), ssDNA probes with high specific activity can be generated (Gyllensten and Erlich 1988). Compared to double-stranded DNA, which is denatured prior to hybridization, these single-stranded probes have the theoretic advantage that re-annealing of probe to the second strand cannot occur. In practice, this does not always occur.

Single stranded antisense RNA probes generated from a specially constructed RNA expression vectors, as originally introduced by Green *et al.* (1983) have also been used (see Chapter 6). 'Antisense' (complementary) RNA with high specific activity can be synthesized using cDNA as a template in the presence of labelled or unlabelled ribonucleotides and RNA polymerase. The advantages of antisense RNA probes over nick-translated DNA probes include a higher specific activity and thermal stability of RNA–RNA hybrids and a defined probe size. All of these factors favour increased sensitivity and consistency of reactions. Additionally, antisense RNA probes do not contain vector sequences, which may cause nonspecific hybridization. Furthermore, competitive hybridization to the complementary strand, which occurs with double stranded probes, is excluded. Finally, and probably most important is the ability to use RNAse to digest unhybridized (single-stranded) probe resulting in extremely low background. All of these advantages result (theoretically) in extremely high sensitivity and excellent signal-to-noise ratios of ISH using cRNA probes.

Antisense oligonucleotides have recently been introduced. These are made by subcloning of synthetic oligonucleotides (20–70 mers) into RNA expression vectors. These probes combine the theoretical advantages of both oligonucleotide and cRNA techniques. Most recently, the synthesis of cRNA probes from synthetic single stranded DNA oligonucleotides was reported: this procedure allows the generation of cRNA probes by synthesis from a bacteriophage promoter and the complementary DNA sequence to be transcribed. The hybridization of the bacteriophage promoter to a complementary promoter sequence and the transcription reaction can be performed simultaneously *in vitro*. This procedure does not require any subcloning or DNA trimming and represents an elegant and most efficient way of probe synthesis (Schlingensiepen and Brysch 1988 and Chapter 1).

Labelling

Two main types of labelling strategy can be used: direct labelling, with direct attachment of the reporter molecule to DNA or RNA; and indirect labelling, in which either a hapten (e.g. biotin or digoxigenin) is attached to the probe and detected by a labelled binding protein (e.g. avidin) or the probe–target hybrid is detected by a specific antibody. Labelling methods for probes currently used are summarized in Table 2.2.

Table 2.2. Labelling of probes for *in situ* hybridization

Isotopic	Non-isotopic
^{3}H, ^{32}P, ^{35}S, ^{14}C, ^{125}I	Alkaline phosphatase
	Biotin, photobiotin
	5-bromodeoxyuridine
	Dinitrophenyl (DNP)
	Digoxigenin
	Ethidium
	Fluororescein
	Luciferase
	Mercury(II)acetate
	N-2-acetylaminofluorene
	N-2-acetylamino-7-iodofluorene
	Sodium metabisulphite
	Sulphonation
	Tetramethylrhodamine
	Antibodies directed to dsDNA, RNA–DNA, and RNA–RNA hybrids

1. *Radioactively labelled probes*, as originally used by Gall and Pardue (1969), are still widely used for ISH, for several reasons: (a) the efficiency of probe synthesis can be monitored more easily; (b) radioisotopes are readily incorporated into the synthesized DNA and RNA using most enzymes; and (c) autoradiography represented a sensitive detection system. Signal detection can be achieved with autoradiography employing liquid emulsions, or with ^{32}P-labelled probes by X-ray films (see Chapter 3). Most commonly used is ^{3}H-labelling because of the high resolution of the autoradiographs. Sections hybridized with ^{3}H-labelled probes, however, usually require long exposure (weeks) for signal detection. If more rapid detection is desired, labelling with high-energy-emitting radioisotopes ^{32}P or ^{35}S can yield autoradiographs within days; ^{14}C- and ^{125}I-labelled probes are not very frequently used for ISH. These isotopes give poor resolution.

2. *Non-isotopic probes*. Problems of safety, waste disposal, and reduced

stability (see Table 2.3) of radioactive probes, and speed of visualization have stimulated development of non-isotopic probes (Burns *et al.* 1985; Bauman 1985; and van der Ploeg *et al.* 1986, and Chapters 10 and 11 for details). Among the choices for non-radioactive labels biotin is currently favoured by many investigators, since immunohistochemistry is well established in most laboratories for the detection of biotin. With the synthesis of biotin-labelled dUTP (Langer *et al.* 1981) the construction of (directly) biotinylated nucleic acids (double-stranded DNA, single-stranded DNA, oligonucleotides, and cRNA) became possible. Allylamino-UTP (Cook *et al.* 1988), photobiotin for DNA (Forster *et al.* 1985) and RNA (Childs *et al.* 1987) labelling, or long-chain diamino compounds covalently linked to biotin (Al-Hakim and Hull 1988) represent modifications of biotinylation of probes. Excellent results with biotinylated synthetic oligonucleotide probes were reported by Guitteny *et al.* (1988). Direct fluorescence-microscopical hybridocytochemistry applying fluorochrome-labelled DNA or RNA (Bauman 1985) is not widely used because of relatively low sensitivity. Developments based on immunohistochemical detection of chemically modified nucleic acids, such as acetylaminofluorene (Landegent *et al.* 1984; Tchen *et al.* 1984), dinitrophenyl (Dnp) groups as a hapten (Shroyer and Nakane 1983), mercurated probes and sulphhydril-hapten ligands (Hopman *et al.* 1986) are promising. Other labels, such as alkaline phosphate, bromodeoxyuridine (Niedobitek *et al.* 1988), and sulphonated probes (Morimoto *et al.* 1987) were introduced recently. The immuno-histochemical localization of DNA–RNA or RNA–RNA hybrids by specific antibodies was described by several groups (Rudkin and Stollar 1974; Raap *et al.* 1984). Digoxigenin substitution will become the reporter of choice (Chapter 12). The merits of radioisotope, biotin, and digoxigenin labelling of probes for ISH are compared in Table 2.3.

Table 2.3. Merits of radiolabelled and other probes

Label	Resolution	Sensitivity	Exposure (days)	Stability (weeks)
^{32}P	+	++	7	0.5
^{35}S	++	+++	10	6
^{3}H	+++	+++	14	> 30
Biotin	+++	++	0.16	> 52
Digoxigenin	+++	+++	0.16	> 52

Hybridization conditions

One of the important advantages of ISH over immunohistochemistry is the fact that the degree of specificity of hybridization reactions can be controlled accurately by varying the reaction conditions. The degree of specificity depends on probe construction, temperature, pH, and forma-mide and salt concentration in the hybridization buffer (see Chapter 1). One has to be aware that despite a number of mismatched bases along the strands of nucleic acids, stable duplexes can be formed under certain conditions of hybridization. The degree of mismatch that can be tolerated in a hybridization reaction is referred to as 'stringency'. Under conditions of high stringency, only probes with high homology to the target sequence will form stable hybrids. At low-stringency (i.e. reactions carried out at low temperature, or in high salt or low formamide concentrations) a probe may bind to sequences with only 70–90 per cent homology, thus resulting in possible non-specific hybridization signals (see below, Chapter 1, and Herrington *et al.* 1990). With radioactively labelled dsDNA or cRNA probes, usually 2–10 ng probe diluted in hybridization buffer is applied per section. The volume of hybridization mixture should be kept as small as possible (i.e. 10–20 μl total volume per section covered with a 22 mm^2 coverslip). When biotinylated probes are used, 10–20 ng DNA probe per section is required. Most hybridization buffers contain a mixture of 50 per cent formamide, $2 \times SSC$ (standard saline citrate: $1 \times SSC = 0.15$ M sodium chloride, 0.015 M sodium citrate). Dextran sulphate (usually 10 per cent) can be added to increase hybridization efficiency (Wahl *et al.* 1979). Depending on desired stringency, and melting temperature of the hybrids, hybridization is undertaken at 37–60 °C. One of the major drawbacks of ISH compared with the hybridization performed on extracts was the limitation in hybridization and washing temperature (and stringency) to temperatures around 50 °C owing to tissue damage and loss of sections from the slides when higher temperatures are used. However, this has been overcome by using paraffin sections on silanized slides (Burns *et al.* 1987, 1988). Optimal annealing temperatures of cDNA and cRNA probes *in situ* are around 50 °C that is 20–25 °C below the melting temperature (Cox *et al.* 1984). Hybridizations employing DNA probes are completed after 2–4 hrs (Lawrence and Singer 1985; Burns *et al.* 1987); reactions with cRNA probes should be incubated overnight. In contrast to the localization of mRNA, hybridization to cellular DNA requires the heating of tissue sections for 5–15 min at 95 °C to denature the target DNA. Washing steps to reduce non-specific binding are performed in decreasing concentrations of SSC (i.e. increasing stringency), usually ending with $0.1 \times SSC$ for the final wash. For ISH employing radiolabelled probes, washing has to be rather extensive (up to several hours), whereas sections hybridized to biotinylated or digoxigenin probes require a short washing procedure (15 min). RNAse treatment of sections after hybridization with antisense RNA probes (DeLeon *et al.* 1983; Lynn *et al.* 1983) to decrease back-ground appears to be more reliable than the comparable S1 nuclease treatment of DNA (Godard 1983). Prior to autoradiography with liquid emulsions, the sections must be air dried (see Chapter 3).

Detailed studies concerning influences of buffers, probe concentration, time and temperature on hybridization efficiency for DNA probes (Lawrence and Singer 1985) and for RNA probes (Cox *et al.* 1984; Höfler *et al.* 1986) have been published. Several improvements have been reported that optimize this technique for different nucleic acid probes and tissue preparation (for review see Coghlan *et al.* 1985; Lawrence *et al.* 1988).

Interpretation of results
Specificity and sensitivity

As mentioned above, low-stringency conditions may result in non-specific reactions. Non-specific cross-hybridization occurs more often with long probes (> 0.5 kb) and with cRNA probes because of their higher affinity to DNA and RNA. Even under stringent reaction conditions, cross-hybridization of probes to related (but not identical) nucleic acid sequences containing high homologous regions (e.g. gene families, such as HPV types 6b and 11) is common (Herrington *et al.* 1990; Chapter 12). Furthermore, non-specific binding to non-related genes and their messages owing to partial sequence homology may occur (Crabbe *et al.* 1985). One of the commonest non-specific reactions in ISH for mRNA localization is binding of the probe to rRNA, which represents more than 90 per cent of total cellular RNA. Therefore, the evaluation of each probe under different conditions of stringency is strongly recommended, and whenever possible each probe should be tested by Northern blot analysis before its application to ISH. The Northern procedure allows the detection of such 'sticky' probes, thus indicating very cautious interpretation of ISH results with these probes. Besides non-specific binding of the probe, the detection system may also cause non-specific results. The widely used biotinylated probes are usually immunohistochemically detected by avidin-binding or the application of anti-biotin antibodies. In several tissues endogenous biotin (vitamin H) may lead to high 'non-specific' signals. Binding of biotin antibodies or avidin to endogenous biotin after ISH procedures cannot be prevented and represents a major restriction of the application of bio-tinylated probes. This is not a problem with digoxigenin-labelled probes (see Chapter 12). Chemography—the occurrence of spurious (non-radio-activity-induced) reaction products in liquid emulsions used for the detection of radiolabelled probes—is caused by the interaction of heavy-metal ions with the emulsion. Minimization or prevention of chemography can be achieved by omission of heavy metals from fixatives; exposure of slides at 4 °C; and observance of constant temperature (15 °C for Kodak NTB2, 20 °C for Ilford K2) during processing procedures.

High sensitivity is one of the advantages of hybridization procedures, and can be monitored exactly in all techniques in which tissue or cell extracts are used (e.g. blot hybridization). Similarly, the sensitivity of ISH in individual cells of homogeneous samples, like cells grown in tissue culture, can be calculated based on the results of hybridizations on cell extracts. Using highly sensitive radiolabelled cRNA probes for mRNA detection, a sensitivity of 20 mRNA copies per cell was reported, which is approximately 10-fold higher compared to ISH employing nick-translated probes (Cox *et al.* 1984; Höfler *et al.* 1987*a*). These results were achieved in experiments under optimized conditions of fixation and hybridization and cannot be transferred to all experiments. Sensitivity of ISH, particularly for mRNA detection, on heterogeneous samples (tissue sections), however, cannot be monitored exactly. When ISH results are evaluated, the varying stability of individual mRNA classes must be considered.

Delay in adequate tissue fixation after (surgical) removal may lead to non-reproducible or even false negative results due to mRNA degradation. Additionally, differences of accessibility of probes are encountered in tissues of different organs. Interestingly, these differences were restricted to oligonucleotide probes and were not observed with cRNA probes (Campell and Habener 1987).

Owing to the higher stability of dsDNA, the problems of loss, degradation, and reduced accessibility of the target sequences are usually not encountered when ISH is used to localize DNA. Furthermore, DNA target sequences on chromosomes are more extended in space and longer (>2 kb) probes can be used. The sensitivity of ISH for DNA localization is therefore much higher than for RNA localization: single-copy genes can be detected in chromosomal preparations and—under certain circumstances—even in tissue sections (see Chapter 1, 4, and 10).

Controls

As with all other histochemical methods, not every positive ISH signal is specific. All available controls, therefore, must be performed in every experiment to prove specificity. Depending on the probe used and the target nucleic acid, several control reactions are available (Table 2.4).

Hybridization of probes to extracted RNA or digested DNA after gel

Table 2.4. Controls for specificity of *in situ* hybridization (ISH)

- Emulsion or non-isotopic detection system controls
- Northern or Southern blot
- Combination of ISH and immunohistochemistry
- Hybridization with different probes complementary to the same target DNA or RNA
- Prehybridization of probe with cDNA or cRNA ('absorption')
- Hybridization with non-specific (vector) sequences, and irrelevant probes
- Pretreatment of sections with RNase or DNase

electrophoresis and transfer on to solid supports (Northern or Southern blot analysis) reveal hybridization to known, defined classes of RNA or DNA only. These methods, comparable to Western blot analysis for specificity controls of antibodies, are important controls to assure probe specificity. Hybridization to positive control sections should reveal the expected distribution of reactivity. If a suitable antibody is available, ISH and immunohistochemistry performed on serial or the same sections (Höfler *et al.* 1987*b*) can confirm that the mRNA is localized in cells containing the peptide product of the same gene. When immunohistochemistry cannot be done, hybridization with probes containing different specific sequences are necessary. Similarly to preabsorbtion of antibodies with specific antigens, prehybridization of probes with specific cDNA or cRNA should yield negative results with ISH. Furthermore, hybridization with non-specific vector sequences, vectors with irrelevant inserts, or sense RNA probes, should produce negative results. Other controls include pretreatment of sections with RNase or DNase to prove that the hybridization depends on the presence of RNA or DNA. Finally, non-specific reactions in the detection system, such as chemography artefacts during autoradiography or non-specific reactions of non-isotopic detection, should be excluded by omitting the labelled probe.

Conclusions and Future Aims

ISH represents a powerful method for specifically localizing DNA or RNA in cells and may therefore provide insight into differences in biosynthetic activity of individual cells. Before commencing an experiment, one has to choose the appropriate system of probe construction, labelling and signal detection. For the localization of DNA and high-copy-number mRNA, dsDNA probes or oligonucleotides, either radiolabelled or non-isotopically labelled, are suitable. In some laboratories, isotopic labelling still represents the method of choice for the generation of probes, particularly cRNA probes, and for the detection of low-copy-number mRNA by ISH. Despite the recent encouraging developments in increasing sensitivity of non-isotopic detection systems, such as biotin antibodies, silver intensification of gold-labelled antibodies (Springall 1984; Burns *et al.* 1985), or reflection-contrast microscopy (Landegent *et al.* 1985), several problems remain to be solved before NISH can be called a 'routine procedure' for use in every laboratory. The development of non-isotopic systems and cloning of new sequences will enable more laboratories to establish this technique and broaden the field of application of NISH. Among the most expanding application areas of NISH is the visualization of pathogenic organisms (Matthews and Kricker 1988 for review), genetics (Hopman *et al.* 1988; Herrington and McGee 1990 for reviews) and oncology (see Chapter 1).

References

Al-Hakim, A. H. and Hull, R. (1988). Chemically synthesized non-radioactive biotinylated long-chain nucleic acid hybridization probes. *Biochem. J.*, **251**, 935–8.

Bauman, J. G. J. (1985). Fluorescence microscopical hybridocytochemistry. *Acta Histochem.*, **31**, 9–18.

Berge-Lefranc, J. L., Cartouzou, G., Bignon, Ch., and Lissitzky, S. (1983). Quantitative *in situ* hybridization of ^{3}H-labeled complementary deoxyribonucleic acid (cDNA) to the messenger ribonucleic acid of thyroglobulin in human thyroid tissues. *J. Clin. Endocrinol. Metabol.*, **57**, 470–6.

Brigati, D. J., Myerson, D., Leary, J. J., Spalholz, B., Travis, S. Z., Fong, S. Z., Hsiung, G. D., and Ward, D. C. (1983). Detection of viral genomes in cultured cells and paraffin-embedded tissue sections using biotin-labeled hybridization probes. *Virology*, **126**, 32–50.

Buongiorno-Nardelli, M. and Amaldi, F. (1969). Autoradiographic detection of molecular hybrids between rRNA and DNA in tissue sections. *Nature*, **225**, 946–7.

Burns, J., Chan, V. T-W., Jonasson, J. A., Fleming, K. A., Taylor, S., and McGee, J. O'D. (1985). A sensitive method for visualising biotinylated probes hybridized in situ: rapid sex determination on intact cells. *J. Clin. Pathol.*, **38**, 1085–92.

Burns, J., Graham, A. K., Frank, C., Fleming, K. A., Evans, M. F., and McGee, J. O'D. (1987). Detection of lowcopy human papillomavirus DNA and mRNA in routine paraffin sections of cervix by non-isotopic *in situ* hybridization. *J. Clin. Pathol.*, **40**, 858–64.

Burns, J., Graham, A. K., and McGee, J. O'D. (1988). Non-isotopic detection of *in situ* nucleic acid in cervix. An updated protocol. *J. Clin. Pathol.*, **41**, 897–9.

Campell, D. J., Habener, J. F. (1987). Cellular localization of angiotensin gene expression in brown adipose tissue and mesentery: Quantification of mRNA abundance using *in situ* hybridization. *Endocrinology*, **121**, 1616–26.

Childs, G. V., Lloyd, J. M., Unabia, G., Gharib, S. D., Wierman, M. E., and Chin, W. W. (1987). Detection of luteinizing hormone β mRNA in individual gonadotropes after castration: Use of a new *in situ* hybridization method with a photo-biotinylated cRNA probe. *Mol. Endocrinol.*, **1**, 926–32.

Coghlan, J. P., Aldred, P., Haralambidis, J., Niall, H. D., Penschow, J. D., and Tregear, G. W. (1985). Hybridization histochemistry. *Analyt. Biochem.*, **149**, 1–28.

Cook, A. F., Voucolo, E., and Brakel, Ch. L. (1988). Synthesis and hybridization of biotinylated oligonucleotides. *Nucl Acids Res.*, **16**(9), 4077–95.

Cox, K. H., DeLeon, D. V., Angerer, L. M., and Angerer, R. C. (1984). Detection of mRNAs in Sea Urchin embroys by *in situ* hybridization using asymmetric RNA probes. *Develop. Biol.* **101**, 485–502.

Crabbe, M. I. (1985). Partial sequence homologies between cytoskeletal proteins, c-myc, Rous sarcoma virus and adenovirus proteins, transducin and beta- and gamma-crystalline. *Biosci. Rep.*, **5**(2), 167–74.

DeLeon, D. V., Cox, K. H., Angerer, L. M., and Angerer, R. C. (1983). Most early-variant histone mRNA is contained in the pronucleus of Sea Urchin eggs. *Develop. Biol.*, **100**, 197–206.

Feinberg, A. P. and Vogelstein, B. (1983). A technique for radiolabeling DNA restriction endonuclease fragments to high specific activity. *Analyt. Biochem.,* **132**, 6–13.

Forster, A. C., McInnes, J. L., Skingle, D. C., and Symons, R. H. (1985). Non-radioactive hybridization probes prepared by the chemical labelling of DNA and RNA with a novel reagent, photobiotin. *Nucl. Acids Res.,* **13**(3), 745.

Gall, G. and Pardue, M. L. (1969). Formation and detection of RNA–DNA hybrid molecules in cytological preparations. *Proc. Natl Acad. Sci.,* **63**, 378–81.

Gall, G. and Pardue, M. L. (1971). Nucleic acid hybridization in cytological preparations. *Methods Enzymol.,* **38**, 470–80.

Godard, C. M. (1983). Improved method for detection of cellular transcripts by *in situ* hybridization. *Histochemistry,* **77**, 123–31.

Green, M. R., Maniatis, T., and Melton, D. A. (1983). Human betaglobin pre-mRNA synthesized *in vitro* is accurately spliced in Xenopus oocyte nuclei. *Cell,* **32**, 681–94.

Guitteny, A.-F., Fouque, B., Mougin, Ch., Teoule, R., and Bloch, B. (1988). Histological detection of mRNAs with biotinylated synthetic oligonucleotide probes. *J. Histochem. Cytochem.,* **36**, 563–71.

Gyllensten, U. B. and Erlich, H. A. (1988). Generation of single stranded DNA by the polymerase chain reaction and its application to direct sequencing of the HLA-DQA locus. *Proc. Natl Acad. Sci.,* 7652–6.

Hafen, E., Levine, M., Garber, R. L., and Gehring, W. J. (1983). An improved *in situ* hybridization method for the detection of cellular RNAs in Drosophila tissue sections and its application for localizing transcripts of the homeotic Antennapedia gene complex. *EMBO J.,* **2**(4), 617–23.

Harper, E. and Saunders, G. F. (1981). Localization of single copy DNA sequences in G-banded human chromosomes by *in situ* hybridization. *Chromosoma,* **83**, 431–9.

Hayashi, S., Gillam, I. C., Delaney, A. D., and Tener, G. M. (1978). Acetylation of chromosome squashes of Drosophila Melanogaster decreases the background in autoradiographs from hybridization with ^{125}I-labelled RNA. *J. Histochem. Cytochem.,* **26**(8), 677–9.

Herrington, C. S. and McGee, J. O'D. (1990). Interphase cytogenetics. *Neurochemical Research.* (In press.)

Herrington, C. S., Burns, J., Graham, A. K., and McGee, J. O'D. (1990). Evaluation of discriminative stringency conditions for the distinction of closely homologous DNA sequences in clinical biopsies. *Histochem. J.* (In press.)

Hopman, A. H. N., Wiegant, J., Duijn, P. van. (1986). A new hybridocytochemical method based on mercurated nucleic acid probes and sulfhydril-hapten ligands. I. Stability of mercury-sulfhydril bond and influence of the ligand structure on immunochemical detection of the hapten. *Histochemistry,* **84**, 169–78.

Hopman, A. H. N., Ramaekers, F. C. S., Raap, A. K., Beck, J. L. M., Devilee, P., van der Ploeg, M., and Vooijs, G. P. (1988). *In situ* hybridization as a tool to study numerical chromosome aberrations in solid bladder tumors. *Histochemistry,* **89**, 307–16.

Höfler, H. (1987). *In situ* Hybridization. A review. *Pathol. Res. Pract.,* **182**, 421–30.

Höfler, H., Childers, H., Montminy, M. R., Lechan, R. M., Goodman, R. H., and Wolfe, H. J. (1986). *In situ* hybridization methods for the detection of somatostatin mRNA in tissue sections using antisense RNA probes. *Histochem. J.*, **18**, 597–604.

Höfler, H., Childers, H., Montminy, M. R., Goodman, R. H., Lechan, R. M., DeLellis, R. A., Tischler, A. S., and Wolfe, H. J. (1987a). Localization of somatostatin mRNA in the gut, pancreas and thyroid gland of the rat using antisense RNA probes for *in situ* hybridization. *Acta. Histochemica. Suppl. XXXIV*, 101–5.

Höfler, H., Ruhri, Ch., Pütz, B., Wirnsberger, G., Klimpfinger, M., and Smolle, J. (1987b). Simultaneous localization of calcitonin mRNA and peptide in a medullary thyroid carcinoma. *Virch. Arch. B.*, **53**(3), 144–51.

Huang, W. M., Gibson, S. J., Facer, P., Gu, J., and Polak, J. M. (1983). Improved section adhesion for immunohistochemistry using high molecular weight polymers of L-lysine as a slide coating. *Histochemistry*, **77**, 275–9.

John, H. L., Birnstiel, M. L., and Jones, K. W. (1969). RNA–DNA hybrids at the cytological level. *Nature*, **223**, 912–13.

Landegent, J. E., Jansen in de Wal, N., Baan, R. A., Hoeymakers, J. H. J., and Van der Ploeg, M. (1984). 2-acetylaminofluorene-modified probes for the indirect hybridochemical detection of specific nucleic acid sequences. *Exp. Cell. Res.*, **153**, 61–72.

Landegent, J. E., Jansen in de Wal, N., Pleom, J. S., and Van der Ploeg, M. (1985). Sensitive detection of hybridocytochemical results by means of reflection contrast microscopy. *J. Histochem. Cytochem.*, **33**(12), 1241–6.

Langer, P. R., Waldrop, A. A., and Ward, D. C. (1981). Enzymatic synthesis of biotin-labeled polynucleotides: Novel nucleic acid affinity probes. *Proc. Natl Acad. Sci.*, **78**(11), 6633–7.

Lawrence, J. B. and Singer, R. H. (1985). Quantitative analysis of *in situ* hybridization methods for the detection of actin gene expression. *Nucl. Acids Res.*, **13**(5), 1777–99.

Lawrence, J., Singer, R. H., Villnave, C. A., Stein, J. L., and Stein, M. (1988). Intracellular distribution of histone mRNAs in human fibroblasts studied by *in situ* hybridization. *Proc. Natl Acad. Sci.*, **85**, 463–7.

Lynn, D. A., Angerer, L. M., Bruskin, A. M., Klein, W. H., and Angerer, R. C. (1983). Localization of a family of mRNAs in a single cell type and its precursors in sea urchin embryos. *Proc. Natl Acad. Sci.*, **80**, 2656–60.

Maniatis, T., Fritsch, E. F., and Sambrock, J. (1982). *Molecular cloning. A laboratory manual.* Cold Spring Harbor Laboratory, New York.

Manuelidis, L., Langer-Safer, P. R., and Ward, D. C. (1982). High-resolution mapping of satellite DNA using biotin-labeled DNA probes. *J. Cell. Biol.*, **95**, 619–25.

Matthews, J. A. and Kricka, L. J. (1988). Analytical strategies for the use of DNA probes. *Analyt. Biochem.*, **169**, 1–25.

McAllister, H. A. and Rock, D. L. (1985). Comparative usefulness of tissue fixatives for *in situ* viral nucleic acid hybridization. *J. Histochem. Cytochem.*, **33**(10), 1026–32.

Morimoto, H., Monden, T., Shimano, T., Higashiyama, M., Tomita, N., Murotani, M., Matsuura, N., Okuda, H., and Mori, T. (1987). Use of sulfonated probes for *in situ* detection of amylase mRNA in formalin-fixed paraffin sections of human pancreas and submaxillary gland. *Lab. Invest.*, **57**(6), 737–41.

Niedobitek, G., Finn, T., Herbst, H., Bornhöft, G., Gerdes, J., and Stein, H. (1988). Detection of viral DNA by *in situ* hybridization using bromodeoxyuridine-labeled DNA probes. *Am. J. Pathol.*, **131**(1), 1–4.

Pardue, M. L. (1986). *In situ* hybridization to study chromosome organization and gene activity. *J. Histochem. Cytochem.*, **34**, 125–6.

Raap, A. K., Marijnen, J. G. J., and van der Ploeg, M. (1984). Anti-DNA.RNA sera. Specificity test and application in quantitative *in situ* hybridization. *Histochemistry*, **81**, 517–20.

Rudkin, G. T., and Stollar, B. D. (1977). High resolution detection of DNA–RNA hybrids *in situ* by indirect fluorescence. *Nature*, **265**, 472–3.

Schlingensiepen, K.-H. and Brysch, W. (1988). Personal communication.

Shroyer, K. P. and Nakane, P. K. J. (1983). Use of DNP-labeled cDNA for *in situ* hybridization. *J. Cell Biol.*, **97**, 377a.

Springall, D. R., Hacker, G. W., Grimelius, L., and Polak, J. M. (1984). The potential of the immunogold-silver staining method for parafin sections. *Histochemistry*, **81**, 603–8.

Tchen, P., Fuchs, R. P. P., Sage, E., and Leng, M. (1984). Chemically modified nucleic acids as immunodetectable probes in hybridization experiments. *Proc. Natl Acad. Sci.*, **81**, 3466–70.

van de Ploeg, M., Landegent, J. E., Hopman, A. H. N., and Raap, A. K. (1986). Non-autoradiographic hybridocytochemistry. *J. Histochem. Cytochem.*, **34**, 126–33.

Varndell, I. M., Polak, J. M., Sikri, K. L., Minth, C. D., Bloom, S. R., and Dixon, J. E. (1984). Visualisation of messenger RNA directing peptide synthesis by *in situ* hybridisation using a novel single-stranded cDNA probe. *Histochemistry*, **81**, 597–601.

Wahl, G. M., Stern, M., and Stark, G. R. (1979). Efficient transfer of large fragments from agarose gels to diazobenzyloxymethyl-paper and rapid hybridization by using dextran sulfate. *Proc. Natl Acad. Sci.*, 76, 3683–7.

3

Radioactive labels: autoradiography and choice of emulsions for *in situ* hybridization

MICHAEL A. W. BRADY AND MARTIN F. FINLAN

Introduction

In situ hybridization is one of the oldest of the modern molecular biology techniques. It was first described in 1969 by Pardue and Gall. The original technique used tritium, which was incorporated into growing cells, and the labelled nucleic acids were detected via exposure to a nuclear track emulsion. An essential feature of *in situ* hybridization is good cellular spatial resolution. To compare the developed silver grains in the emulsion with the other cellular features (which may be stained) only β-emitting isotopes can be used. The photographic detection system (nuclear track emulsion) was developed for high-energy physics in the 1950s. The basics of emulsion detection technology have changed little since then, but formulation of the emulsion has been improved to make it more suitable for *in situ* hybridization. Though *in situ* hybridization has been improved over the years, it still takes days or weeks to obtain a result with radioactive probes compared to hours with non-radioactive probes.

The potential of other labels such as ^{125}I and ^{32}P for *in situ* hybridization was recognized in the early 1970s. Interestingly, in 1973, the potential of ^{125}I for *in situ* hybridization was debated by W Prensky *et al.* (1973) while even today the potential of this isotope is still being explored (Lewis *et al.* 1986; Allen *et al.* 1987).

The limitations of radioisotopic detection for *in situ* hybridization were recognized from an early stage and the first use of a non-radioactive alternative was described by Manning *et al.* in 1975. Biotin was covalently attached to the nucleic acid via cytochrome *c*; this was detected, post hybridization, by visualizing avidin microspheres at the electron-micro-scopical level. In recent times a number of non-radioactive alternative technologies have been developed; these are reviewed in Chapter 4.

Today *in vitro* biochemical labelling of purified nucleic acids is univer-sally employed. These methods provide high efficiency and flexibility for radioactive labelling of either DNA or RNA probes. In the future, it seems likely that chemically synthesized oligonucleotides will be increasingly used. Recent progress in the preparation and application of stable oligonucleotide RNA probes could be especially interesting (Lamon *et al.* 1989). Significant advances have recently been made by the Caruthers group (Neilson *et al.* 1988 and Wolfgang *et al.* 1988) in the synthesis of DNA with phosphorodithioate linkage. Such probes seem extremely stable and are of obvious interest for *in situ* hybridization. Proper precautions for the safe handling of radioisotopes need to be followed.

The combination of radioisotopes and detection via a contact emulsion has been successfully used for *in situ* hybridization for many years. It has been applied widely in biomedical research and increasingly in pathology. Some still regard it as the 'gold standard' (but see Bhatt *et al.* 1988 and Syrjanen *et al.* 1988). Success in the use of this method is dependent upon a number of factors (see below). These factors require some compromises for good *in situ* hybridization results and the researcher needs to be aware of the underlying principles of detection to select the best combination of radiolabel and assay conditions to suit the particular experiment.

Radioactive labels

The requirements of *in situ* hybridization place severe limitations on the radioisotope that can be used. Any label, whether it is radioactive or not, needs to mimic as closely as possible the natural molecule; the label is chosen with this in mind, viz. to be compatible with cellular or subcellular detection of the hybrid, and to provide rapid and reliable results with a convenient assay format. To meet all of these objectives would require a 'natural' nucleotide radioisotope (of half-life between 3 and 15 days) with a high specific activity and which can readily be incorporated into nucleotides in an efficient and economic way. It would be a pure β-emitter with an energy of around 0.04 Mev. This ideal radioisotope would allow a rapid assay result (~ 1 day or less) with excellent localization and the minimum of safety problems. Unfortunately, none of the isotopes available combines these characteristics and hence the design of the *in situ* hybridization procedure is a compromise between the properties of available radioisotopes, the thickness of the sample and emulsion, and the type of assay result sought.

Table 3.1 gives the properties of the various isotopes that have been employed successfully for *in situ* hybridization. The properties of each of these radioisotopes is briefly described.

Phosphorus 32

The ready availability of this isotope and labelled nucleotides has been one of the major developments that has underpinned the progress of modern

Table 3.1. Characteristics of radionuclides used in *in situ* hybridization

Radionuclide	$T_{1/2}$*	Type/max. energy of emission (MeV)	Specific activity range of nucleotides (TBq)**	Labelling methods	Typical specific activity of probe (dpm/μg)	Detection limit† (dpm/cm^2)
^{32}P	14.3d	β/1.71	15–222 (400–6000)	Nick translation, random priming, *in vitro* transcription, end labelling	5×10^8 5×10^9 1.3×10^9 5×10^6	50
^{33}P	28d	β/0.25	15–100 (400–2500)	Nick translation, random priming, *in vitro* transcription, end labelling	1×10^8 7×10^8 1.3×10^9	300
^{35}S	87.4d	β/0.167	15–55 (400–1500)	Nick translation, random priming, *in vitro* transcription, end labelling	1×10^8 7×10^8 1.3×10^9	400
^{125}I	60d	β/0.035	40–80 (1000–2000)	Nick translation, random priming, direct iodination	1×10^8 1.5×10^9 2×10^8	100
^{3}H	12.35y	β/0.018	0.1–4 (25–100)	Nick translation, random priming, *in vitro* transcription, end labelling	5×10^7 1.5×10^8 5×10	8000‡

* Radionuclide half-life.
** Values in parentheses are equivalents in curies.
† Detection using an intensifying screen.
‡ Detection by fluorography using Amplify.

molecular biology. The method of production in the reactor is ^{32}S (np) $\rightarrow$ ^{32}P and its short half-life results in high specific activities of phosphate, up to ~ 220 TBq/mmol (6000 Ci/mmol) close to the theoretical maximum of ~ 333 TBq/mmol (~ 9000 Ci/mmol). This allows a high proportion of the phosphate groups in the nucleic acid probe to be labelled. In practice, this number is limited by bond breakage and progressive degradation of the labelled nucleic acid, though this may be less important for *in situ* hybridiz- ation where degradation of the nucleic acid is not important post-hybridiz- ation (except in quantitative experiments). Probe preparation methods for labelling with ^{32}P are well established (Table 3.1 and Appendix). The 1.7 MeV β energy is too penetrating for good *in situ* hybridization resolu- tion but nevertheless allows a rapid assay where this is required.

Phosphorus 33

^{33}P is an attractive radioisotope, particularly from an *in situ* hybridization aspect because of its lower β energy and slightly longer half-life. However it is only available in limited quantities and at a price premium. The major difficulty is due to its production route in the reactor, ^{33}S (np) $\rightarrow$ ^{33}P. The stable ^{33}S starting material is difficult to separate owing to its low natural abundance of 0.7 per cent. Further, the presence of small quantities of ^{32}S (95 per cent natural abundance) lead to co-production of ^{32}P in the reactor as an unwanted contaminant. Currently, Russia is producing larger quantities of ^{33}P, but it still remains an expensive specialist isotope. Spatial resolution using this isotope is much improved owing to the lower β energy. Although of similar β energy to ^{35}S, ^{33}P has the advantage of being a 'natural' nucleotide label and should be theoretically better than ^{35}S (see below).

Sulphur 35

^{35}S is a comparative newcomer for nucleic acid labelling; its major applica- tion being the improvement of DNA sequencing (Biggin *et al.* 1983). Further applications have been reviewed recently (Eckstein and Gish 1989). A number of *in situ* hybridization studies showing excellent results have now been reported (Harper *et al.* 1986; Chayt *et al.* 1986; Stoley *et al.* 1986). Again, the method of production, ^{35}Cl (np) $\rightarrow$ ^{35}S, ensures almost carrier-free radioisotopes, i.e. high specific activity. The β energy is more attractive for the *in situ* hybridization but is still rather penetrating. So far, the wide utility and availability of ^{35}S-labelled nucleotides has been limited by difficult preparation chemistry. New preparation procedures may improve this situation (Ludwig *et al.* 1989). It needs to be borne in mind that the replacement of oxygen by sulphur in phosphate groups may result in altered chemical or biochemical properties of both nucleotide triphos- phate and nucleic acid and this may lead to unwanted results, e.g. higher background (Brandtlow *et al.* 1987). Special precautions, e.g. DTT in washes, are needed to minimize this.

Iodine 125

The labelling of nucleic acids with ^{125}I was developed many years ago (Heiniger *et al.* 1973). This isotope will only label pyrimidines, cytosine, uridine, and uracil (either ribose or deoxyribose) where it is added to the 5' ring position with seemingly little effect on hybridization. The interest for *in situ* hybridization derives from electrons emitted with an energy of approximately 0.004 MeV. This gives good spatial resolution and is analogous to high-specific-activity tritium. The application of ^{125}I has been extended to RNA probes (Allen *et al.* 1987), though the general use of this isotope for *in situ* hybridization has perhaps been limited by the availability of labelled nucleotides.

Tritium

Tritium is a well established isotope for *in situ* hybridization. It gives superb resolution and is still widely used, in particular where definitive cellular resolution is sought (Ayer-Lelievre *et al.* 1988). The classical drawback to the use of tritium is the low β energy and the long exposure times that are required to obtain good results. Although production methods to produce increased specific activity nucleotides have been successful, counting times to obtain results are still measured in weeks and not days.

As we have noted, *in vitro* biochemical labelling methods to prepare the radiolabelled nucleic acids have been developed in recent years, including end labelling, nick translation, random priming and *in vitro* transcription in the presence of labelled nucleotides. These are now standard procedures in many laboratories and are dealt with in more detail in the Appendix. Appropriate precautions must be taken when handling radioisotopes. The presence of radioisotopes can usually be detected by sensitive radiation monitoring and the researcher's attention is drawn to the safety precautions in the Appendix.

Detection of beta particles in autoradiographic emulsion

The detection of beta particles in an emulsion is due to the ionization that occurs from the passage of fast electrons (β particles) in matter; a large excess of energy is deposited locally from each interaction of a fast electron with the atoms in the emulsion; this energy causes the reduction of Ag^+ ions to metallic silver, which subsequently aggregate to form a latent image. The latent image is subsequently developed and fixed by normal photographic procedures. The emulsion will faithfully record the presence of a single electron provided it interacts with the emulsion and not the under-lying glass slide. Because each interaction results in large energy loss, at least one grain will result. This contrasts with the interaction of light photons with photographic emulsion, where there is a rate effect, and well known 'low-light reciprocity failure' which requires a minimum number of photons per unit time before the emulsion records the presence of the photons.

The detection of β particles in a nuclear track emulsion is a complex process that depends upon a number of factors: energy/range of the β particles, spectrum of the β-emitter, sample and emulsion thickness, exposure time plus the basic properties of the emulsion. Each of these topics is briefly outlined to illustrate the principles, to guide the researcher as to the choices available and to delineate the important parameters that need to be controlled for a successful *in situ* hybridization result based on radioactive detection (see also Chapter 2).

Range and spectrum of β particles

The range of β particles of the isotopes commonly used in *in situ* hybridization is shown in Table 3.2. This table shows that the range of tritium β emissions is somewhat short for microscopical analysis, while the range of most of the other isotopes is too long. Worth noting is the reduced range of β particles in the nuclear track emulsion, which derives mainly from the increased density of the dried emulsion as a result of the high loading of silver bromide. In reality not all of the water is removed by drying, and hence the actual density of the dried emulsion may be variable. The situation is further complicated by the continuous nature of the energy spectrum of the emitted β particles. (Figures 3.1 and 3.2 show examples.)

This results from the basic nuclear decay properties of a β emitter. A neutrino, an uncharged particle of very low mass, is generated at the same time as the β particle. The neutrino carries the balance of the energy lost from the nuclear decay via a spin state. Hence, though there is a determined amount of energy lost in each nuclear decay, the practical end result is that

Table 3.2. Range of beta particles

Radionuclide/energy (MeV)	Maximum range in tissue (μm)	Maximum range in emulsion* (μm)
^{32}P/1.71	8000	2000
^{33}P/0.25	600	150
^{35}S/0.167	300	80
^{125}I/0.035 γ	36 000	~ 9000
β/0.035 (max)	20	5
β/0.004	0.2	0.05
^{3}H/0.018	5	1

* Assumes all water is removed.

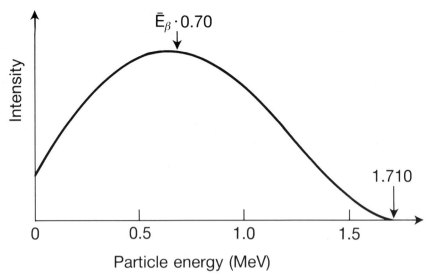

Fig. 3.1. β Spectrum of ^{32}P.

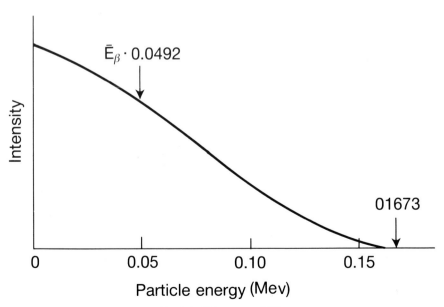

Fig. 3.2. β Spectrum of ^{35}S.

β particles are emitted with a range of energies from a few electron-volts up to the maximum of the nuclear decay with neutrinos making up the difference in energy balance. This results in the characteristic β energy spectrum such as those noted above. Since β-emitters produce particles of such a wide energy range, a rule of thumb for treating this is to assume that the β particle is emitted as a monoenergetic particle with 1/3 of the maximum energy. The range of the β particle in tissue, cells, and emulsion needs to be reduced from the maximum ranges shown in Table 3.2 to take due account of the spectrum of the β-emitter. The neutrino, because of its lack of charge and low mass plays no part in the detection process in the emulsion. Once the electrons from a typical β-emitter have travelled even small distances in the emulsion, the energy spectrum may be modified as indicated in Fig. 3.3.

In addition to the β spectrum we need to note a further complication, viz. that the amount of energy lost per unit length by the electron is not constant as it passes through matter. As a fast electron slows towards the end of its track, the number of collisions per unit length increases until it is finally arrested. In practical terms this results in increased numbers of grains towards the end of a β particle path in an emulsion. This effect is superimposed on the tortuous track appearance of fast electron in an emulsion. Unlike heavier charged particles which follow a straight track with little deviation, the β particle, owing to its very small mass, can undergo massive nuclear and electron scattering as it interacts, often resulting in major deviation and hence uncertainty as to its spatial point of origin. Backscattering of more energetic electrons between the glass microscope slide,

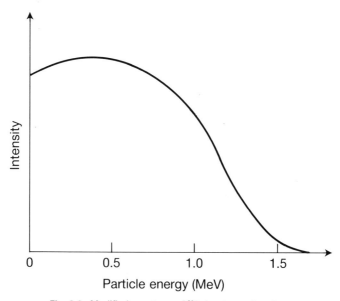

Fig. 3.3. Modified spectrum of ^{32}P due to an absorber.

the sample, and emulsion, and from the top surface of the emulsion can also influence the background and the final result.

The electrons from the decay of ^{125}I are different from those of the normal β emitter. They are mainly of nuclear origin and result from both electron capture and internal conversion. This results in the emission of both electromagnetic radiation, 35 keV, plus tellurium K X-rays together with β particles with a range of energies up to 35 keV. Approximately 2.6 electrons are produced from each nuclear decay and the majority of these have an energy of 3–4 keV. As Table 3.2 shows, the electromagnetic radiation, in comparison to the β radiation, is very penetrating and can give rise to background grains in the emulsion.

Because of the factors briefly outlined above, a theoretical treatment of β particle detection in nuclear track emulsion is not really possible and hence a pragmatic approach is generally adopted.

Thickness of sample and emulsion

From the considerations in the previous section is it clear that a combination of radiolabel, sample, and emulsion thickness can be chosen to yield either a high-sensitivity result with relatively poor resolution, or alternatively a high-resolution result with the sacrifice of time. For a detailed treatment of this, the reader is referred to the work of Rogers 1979.

It is commonly accepted that good cellular resolution can only be obtained with emulsion thicknesses of less than 5 μm. As the emulsion thickness increases beyond this, the contribution from β particles scattered at wider angles becomes more evident. This leads to an increase in background and a lack of precision as to the origin of the β particles and hence uncertainty as to the cellular localization of the hybrid. Thicknesses of 3–5 μm can be achieved by dipping the prepared sample in a diluted emulsion mix (1:1) and allowing it to drain. Dipping slides into neat emulsion as supplied by some manufacturers results in a coating of ~10 μm. Although coating slides by dipping is now widely used, it does not result in an even emulsion coat. The use of stripping film will result in a uniform emulsion layer, but is not as convenient as dipping and hence is less favoured today.

There are a number of emulsion types available, e.g. Ilford G5, K5, L4 and Kodak NTB2. While all are sensitive to charged particles, not all of them are suitable for optical microscopy; for example, the Ilford K range has been specially formulated for optical microscopy. Amersham has recently introduced a convenient ready-to-use emulsion, LM-1 for light microscopy, specially formulated for *in situ* hybridization. In general, small crystal sizes of silver bromide result in a more efficient emulsion. The different emulsion types offer a range of crystal sizes and sensitivities and the researcher needs to consult the manufacturer's recommendations before selecting an emulsion that suits the particular assay and choice of radioactive label. Appropriate care is needed in the handling of emulsions. The usual precautions for photographic materials are required, and additionally care is needed to avoid air bubbles, cracking, or defects of the emulsion surface. For long exposures, emulsions need to be stored dry in the dark and, if more energetic radionucleotides are being used, stored, in such a way that the scattered radiation from one slide does not contribute to the background of adjacent slides. Detailed procedures for slide preparation and handling are contained in the Appendix. As with radioisotopes, the choice of emulsions is something of a compromise. A very sensitive emulsion may give rise to higher background from naturally occurring radiation and may not be suitable for the longer exposures required for tritium.

Special care is needed to obtain a good signal if tritium or ^{125}I are being used. For example, with a uniformly labelled tritium cell or tissue, only the top few micrometres in contact with the emulsion will result in any measurable signal. Alternatively, if the tritium labelled probe is localized to DNA or RNA in the cell nucleus this can be several micrometres below the surface of the sample on the slide, and virtually no signal will be detected. It has been suggested that the liquid emulsion may creep into the surface of the sample, e.g. into surface imperfection or pores; this would result in a more intimate contact between the emulsion and the radioactive label than might otherwise have been the case. There is no published data on this.

Computer simulated imaging of in situ hybridization detection

In an attempt to understand and visualize the factors that influence signal detection in an *in situ* hybridization experiment, a special computer program was generated. This program prints out a dot for each interaction of a primary or secondary electron with surrounding nuclei or electrons, using known values; proper account is taken of the energy loss from each electron interaction and the program results in a 'scattergram' simulating the deposition of grains in an emulsion, in which both the distribution and the number of dots are recorded.

Imaging of a β-emitter by autoradiography, in a tissue section, leads to grains in the developed emulsion that are clearly much larger in area than the actual interaction of the β-emitter itself. Autoradiography thus leads to lower resolution than would be the case if the electrons travelled only orthogonally with respect to the emulsion. The problem has been tackled mathematically and latterly by computer, with Monte Carlo codes using known scattering cross sections. The computer-modelling program simulates the behaviour of the electrons in an iterative fashion to try to optimize the environment of the emitter to decipher spatial resolution.

The program considers the β-emitter to be distributed along a plane shown in section (Fig. 3.4), emitting randomly in time and space. The angle through which the electron is scattered at each collision is derived from the approximate equation (Hine and Brownell 1956):

$$\theta^2 = \frac{4\pi NZ(Z+1)e^4 t \ \log\ [4\pi ZNt(\hbar/m_0 v)^2]}{p^2 v^2}$$

θ^2 is the mean square deflection of the electrons; t is the absorber thickness in g/mm^2; p is the electron momentum; Z is the atomic number of the absorber atoms; $\hbar$ is h (Planck's constant)/2π, m_0 is the rest mass of the electron; v is the electron velocity; and N is the number of atoms/unit volume.

After travelling a distance that is randomly determined between limits, the electron is deemed to be scattered, and a dot is printed as projected back onto the emitting plane. The new direction of the electron is randomly chosen in free space, and the process is continued until the electron comes to rest. The distance travelled in successive scattering is progressively reduced. This process is repeated for discrete energies chosen roughly to mimic the energy spectrum of the β source. The scattering points are assumed to produce a latent image in a grain of the silver halide, and a histogram is plotted that represents the distribution of the intensity of the eventual image.

The program was run to study the effect of different parameters, such as β energy, emulsion thickness, and scattering of material that surrounds the sample and emulsion. In addition to printing a dot for each electron interaction, the program prints both a histogram about the line activity source of the electrons interacting in the emulsion, and the total number of electrons interacting in the emulsion in the given period. This allows us to simulate

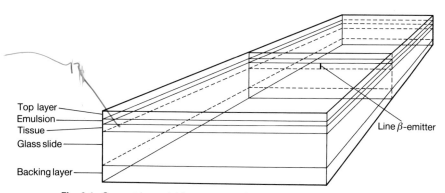

Fig. 3.4. Geometric model for computer-simulated imaging of a β-emitter.

the resolution of the autoradiographic image. Two examples illustrate the kind of result obtainable from this program. In the illustrations Figs. 3.5 and 3.6, the effect of increasing the emulsion thickness from 2 to 10 μm can be seen.

Figure 3.7 shows the effect of increasing the atomic number, and density of the material in contact above the emulsion layer and below the sample layer. If the latter is increased to 6.4 g/cm^3 (lead glass) the improved distribution of electron interaction about the line source is evident despite the increased emulsion thickness.

In practical terms, it is not always easy to realize the advantage that can result from altering the experimental design as suggested above; for example, it is not easy to increase the emulsion thickness without introducing defects into the emulsion that would override any likely improvement. In brief, these computer studies have served to confirm what Rogers in his book refers to as 'the conflicting demands of resolution and efficiency' (Rogers 1979).

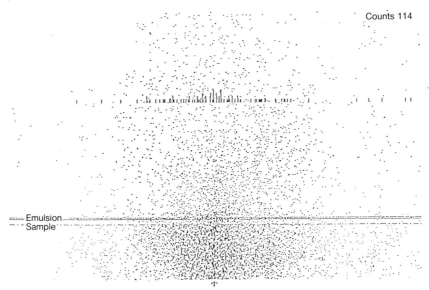

Counts 114

Emulsion
Sample

Fig. 3.5. Computer-simulated autoradiography. Glass density, 2.4 g/cm^3. Glass thickness, 1000 μm. Emulsion thickness, 2 μm. Sample thickness, 10 μm. Top layer density, 1 g/cm^3. Isotope, ^{35}S. Activity, 86 pCi for 60 s.

Examples of *in situ* hybridization with commonly used radioisotopes

Figures 3.8, 3.9, and 3.10 show the typical results that can be obtained on either cells or tissues using the isotopes commonly used for *in situ* hybridization. Samples are typically 10 μm thick with an emulsion coat of around

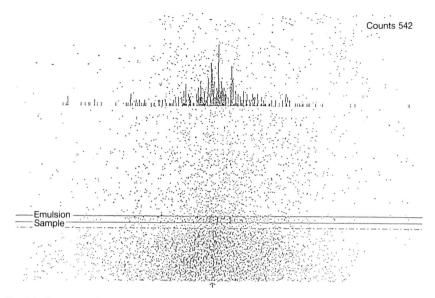

Counts 542

Fig. 3.6. Computer-simulated autoradiography. Glass density, 2.4 g/cm³. Glass thickness, 1000 μm. Emulsion thickness, 10 μm. Sample thickness, 10 μm. Top layer density, 1 g/cm³. Isotope, ³⁵S. Activity, 86 pCi for 60 s.

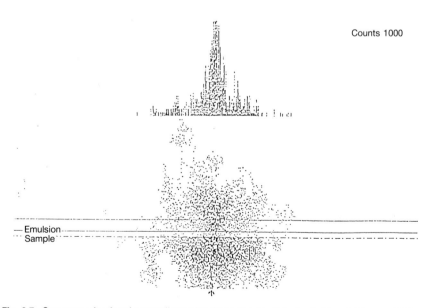

Counts 1000

Fig. 3.7. Computer-simulated autoradiography. Glass density, 6.4 g/cm³. Glass thickness, 1000 μm. Emulsion thickness, 20 μm. Sample thickness, 10 μm. Top layer density, 6.4 g/cm³. Isotope, ³⁵S. Activity, 86 pCi for 60 s.

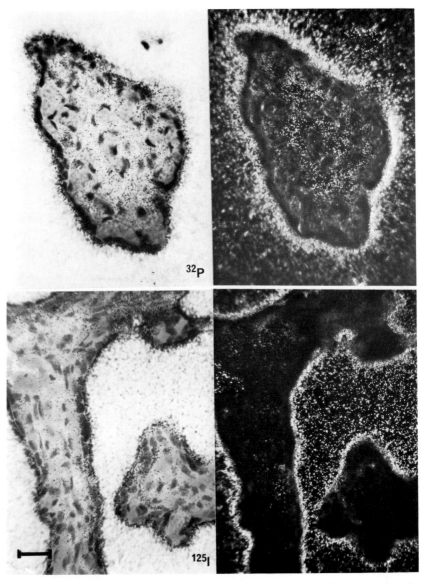

Fig. 3.8. Cryostat sections of human placenta fixed in 2% glutaraldehyde. The sections have been hybridized with human placental lactogen RNA probes labelled with either ^{32}P (specific activity 1.3×10^9 dpm/μg) or ^{125}I (specific activity 2.5×10^9 dpm/μg). The placenta is counterstained with haematoxylin/eosin and shows strong hybridization signal in the syncytial trophoblast surrounding the chorionic villi. The sections are shown in bright field (left) and dark field (right). Autoradiography exposure times: ^{32}P, 4 days; ^{125}I, 11 days. Scale bar = 20 μm.

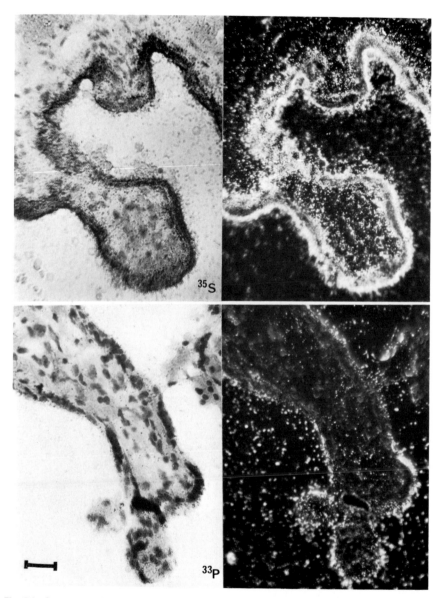

Fig. 3.9. Cryostat sections of human placenta fixed in 2% glutaraldehyde. The sections have been hybridized with human placental lactogen RNA probes labelled with ^{35}S (specific activity 6.6×10^8 dpm/μg) or ^{33}P (specific activity 5.5×10^8 dpm/μg). The placenta is counterstained with haematoxylin/eosin and shows strong hybridization signal in the syncytial trophoblast surrounding the chorionic villi. The sections are shown in bright field (left) and dark field (right). Autoradiography exposure times: ^{35}S, 6 days; ^{33}P, 8 days. Scale bar = 20 μm.

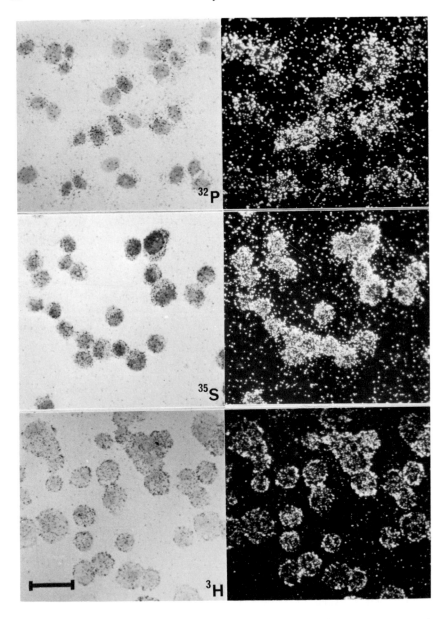

Fig. 3.10. Autoradiography of cultured chicken embryo cells that have been cytospun on to glass slides, fixed in Bouin's fixative, and hybridized with chicken β-actin RNA probes. The autoradiographs demonstrate the degree of resolution produced by different radioactive labels. The cells have been hybridized with a ^{32}P-labelled RNA probe (specific activity 1.3×10^9 dpm/μg), exposed for 4 days; a ^{35}S-labelled RNA probe (specific activity 6.6×10^8 dpm/μg) exposed for 6 days; and a ^{3}H-labelled RNA probe (specific activity 8.6×10^7 dpm/μg) exposed for 14 days. The preparations are shown in bright field (left) and dark field (right). Scale bar = 10 μm.

5 μm. The signal and background can be compared by means of the bright field and dark field exposures. These examples illustrate many of the principles discussed above and underline that some understanding of the principles of β particle autoradiography is required to enable the researcher or 'assayist' to make the right combination of choices to suit the particular *in situ* hybridization question.

Acknowledgement

The examples of *in situ* hybridization in this chapter were kindly prepared by Amersham's *In-Situ* Hybridization Group in Corporate Research and their help is gratefully acknowledged.

References

Allen, J. M., Sasek, C. A., Martin, J. B., and Heinrich, G. (1987). Use of complementary ^{125}I labelled RNA for single cell resolution by *in situ* hybridization. *Biotechniques*, **5**(8), 774–7.

Ayer-Lelievre, C., Olson, L., Ebendal, I., Seiger, A., and Peterson, H. (1988). Expression of the beta nerve growth factor gene in hippocampal neurons. *Science*, **240** (June 3), 1339–41.

Bhatt, B., Burns, J., Flannery, D., and McGee, J. O'D. (1988). Direct visualization of single copy genes on banded metaphase chromosomes by non-isotopic in situ hybridisation. *Nucl. Acids Res.*, **16**, 3951–61.

Biggin, M. D., Gibson, T. J., and Hong, G. F. (1983). Buffer gradient gels and its ^{35}S label as an aid to rapid DNA sequence determination. *Proc. Natl Acad. Sci. USA*, **80**, 3963–5.

Brandtlow, C. E., Heumann, R., Schwab, E., and Thoenen, H. (1987). Cellular localization of nerve growth factor synthesis by *in situ* hybridization. *EMBO J.*, **6**(4), 891–9.

Chayt, K. S. *et al.* (1986). Detection of HTLV-III RNA in lungs of patients with AIDS and pulmonary involvement. *J. Amer. Med. Assoc.*, **256**(17), 2356–9.

Eckstein, F. and Gish, G. (1989). Phosphorothioates in molecular biology. *TIBS*, **14**(3), 97–100.

Harper, M. E., Marselle, L. M., Gallo, R. C., and W-Staal, F. (1986). Detection of lymphocytes expressing human T-lymphotropic virus type III in lymph nodes and peripheral blood from infected individuals by *in situ* hybridization. *Proc. Natl Acad. Sci. USA.*, **83**, 772–6.

Heiniger, H. J., Chen, H. W., and Commerford, S. L. (1973). Iodination of ribosomal RNA *in vitro*. *Int. J. Appl. Rad. Isotopes*, **24**, 425–7.

Hine, G. J. and Brownell, G. L. (1956). *Radiation dosimetry*, Chap. 2. Academic Press, London.

Lewis, M. E., Arentzen, R., and Baldino, F. Jr. (1986). Rapid, high resolution *in situ* hybridization. Histochemistry with radioionated synthetic oligonucleotides. *J. Neurosci. Res.*, **16**, 117–24.

Lamond, A. I., Sproat, B. S., Ryder, U., and Hamm, J. (1989). Probing the structure and function of U2 snRNP with anti-sense oligonucleotides made of 2'-OMe RNA. *Cell*, **58**(2), 383–90.

Ludwig, J. and Eckstein, F. (1989). Rapid and efficient synthesis of nucleoside 5'-O-(1-thiotriphosphates), 5'-triphosphates and 2',3'-cyclophosphorothioates using 2-chloro-4H-1,3,2-benzodioxaphosphorin-4-one. *J. Organic Chem.*, **54**, 631–5.

Manning, J. E., Hershey, N. D., Broker, T. R., Pellegrini, M., Mitchell, H. K., and Davidson, N. (1975). A new method of *in situ* hybridization. *Chromosoma*, **53**, 107–17.

Pardue, M. L. and Gall, J. G. (1969). Molecular hybridization of radioactive DNA to the DNA of cytological preparations. *Proc. Natl Acad. Sci. USA*, **64**, 600–4.

Prensky, W., Steffensen, D. M., and Hughes, W. L. (1973). The use of iodinated RNA for gene localization. *Proc. Natl Acad. Sci. USA*, **70**(6), 1860–4.

Rogers, A. W. (1979). *Techniques of autoradiography*, (3rd edn), Chaps, 3, 4, 6. Elsevier/North Holland Biomedical Press.

Sproat, B., Lamond, A. I., Beijer, B., Neuner, P., and Ryder, U. (1989). Highly efficient chemical synthesis of 2'-O-methyl oligoribonucleotides and tetra-biotinylated derivatives—novel probes that are resistant to degradation by RNA or DNA specific nucleases. *Nucl. Acids Res.*, **17**(9), 3373–86.

Stoley, M. H. *et al.* (1986). Human T-cell lymphotropic virus type III infection of the central nervous system. A preliminary *in situ* analysis. *J. Amer. Med. Assoc.*, **235**(17), 2360–4.

Syrjanen, S., Partanen, P., Mantyjarvi, R., and Syrjanen, K. (1988). Sensitivity of in-situ hybridisation techniques using biotin and ^{35}S labelled human papilloma virus (HPV) DNA probes. *J. Virol. Methods*, **19**, 225–38.

Wolfgang, K., Brill, D., Neilsen, J., and Caruthers, M. H. (1988). Synthesis of dinucleoside phosphorodithioates via thioamidites. *Tetrahedron letters*, **29**(43), 5517–20.

Appendix 3.1
Methods of radioactively labelling nucleic acid probes

The choice of which type of probe to use and the labelling procedure is simplified by the following two tables (Tables A3.1 and A3.2). The first table summarizes the overall choice available including the origin of the probe, while the second outlines the advantages and disadvantages of the various types of probes for *in situ* hybridization.

Nick translation

One of the oldest and most popular techniques for the labelling of DNA is nick translation (Rigby *et al.* 1977). This uses the action of *E. coli* DNA polymerase I (DNA pol I) to extend a single stranded nick in double stranded DNA in $5' \rightarrow 3'$ direction. The nicks are introduced by the simultaneous action of pancreatic deoxyribonuclease I (DNase I). If one or more appropriately radioactively labelled deoxynucleoside triphosphates is introduced into the reaction mixture, radioactively labelled DNA results. Both the quality and quantity of DNase I is crucial to obtain reliable results with a uniformly labelled population of DNA molecules. For *in situ* hybridization probe lengths of a few hundred bases are preferred. It is important

Table A3.1. Selection of probes for *in situ* hybridization*

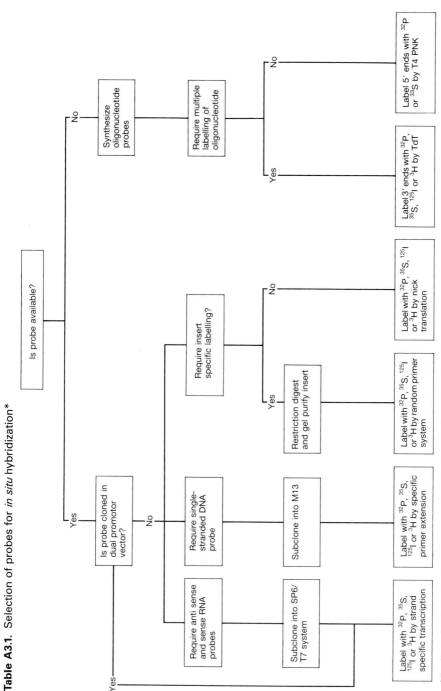

*Radioactive labels can be replaced by biotin, etc. (see Chapter 4).

Table A3.2. Advantages and disadvantages of probes for *in situ* hybridization

	Advantages	Disadvantages
RNA probes	RNA:RNA hybrids have high stability. No probe denaturation needed. No reannealing during hybridization. Probe is strand specific—may be anti-sense or sense probe. Probe is free of vector sequence. Template removal is easy. Post hybridization RNAase treatment removes non-hybridized probe.	Subcloning of probe into dual promoter vector required. Care required to avoid RNase degradation of probe. Narrow optimum hybridization temperature with some probes.
Double strand DNA probes	No subcloning required. Choice of labelling methods available. Hybridization temperature less critical. Possibility of amplification of signal by networking using vector sequences.	Probe denaturation required. Reannealing in hybridization reaction. Hybrids less stable than with RNA probes. Gel purification necessary to remove vector sequences.
Single strand (M13) DNA	Probe free of vector sequence. No probe denaturation required. No reannealing during hybridization.	Subcloning into M13 required. Hybrids less stable than RNA probes. Technically cumbersome.
Oligonucleotide probes	No cloning required. No self hybridization. Good target penetration. Can be constructed to deduced sequence from amino acid data. Can avoid homology with related sequences. Stable.	Limited labelling methods. Small size limits amount of label carried. Subject to 'design' errors. Access to oligonucleotide synthesiser needed. Hybrids less stable than RNA probes. Only short sequence available for hybridization.

that pure double-stranded DNA is used if good results are to be obtained. 0.1–1 μg of DNA is typically used in a labelling reaction. A number of commercial kits are available with detailed protocols to obtain probes of the required quality. Typical results are illustrated in Figs A3.1 and A3.2 using ^{32}P-labelled deoxynucleoside triphosphates. In addition to radioactively labelled nucleotides, non-radioactive molecules, e.g. biotin and digoxigenin-labelled nucleotides, can also be introduced via this technique.

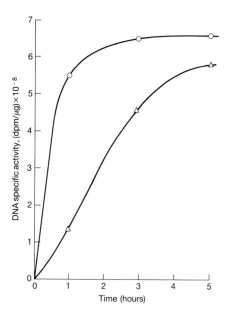

Fig. A3.1. Time courses of nick translation reactions involving $(\alpha$-^{32}P)dCTP at two different specific activities. Reaction mixtures (30 μl) contained 0.3 μg λDNA restriction fragments, 60 picograms DNase I, 3 units DNA polymerase I and a twofold excess of each unlabelled dNTP (dATP, dGTP, dTTP) over $(\alpha$-^{32}P)dCTP whose concentration was either ∼2 μM (O) (specific activity ∼2000–3000 Ci/mmol) or 10 μM (△) (specific activity 400 Ci/mmol).

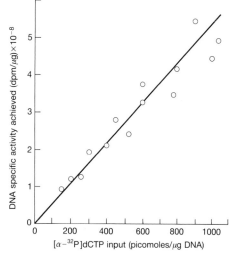

Fig. A3.2. Relationship between $(\alpha$-^{32}P)dCTP input and DNA specific activity achieved in nick translation. Reaction mixtures (100 μl) contained 0.5–1 μg λDNA, 50–200 picograms DNase I, 5–10 units DNA polymerase I and a twofold excess of each unlabelled dNTP (dATP, dGTP, dTTP) over $(\alpha$-^{32}P)dCTP (specific activity 400 Ci/mmol); each reaction was allowed to proceed until net incorporation of labelled nucleotide had ceased. The DNA specific activity was calculated from the percentage of input radioactivity incorporated.

RNA polymerase

These methods are based mainly on the use of bacteriophage RNA polymerases, most commonly SP6 RNA polymerase and T7 RNA polymerase. The enzyme copies the DNA template from a specific promoter to generate a series of radioactively labelled RNA transcripts. The RNA polymerase incorporates the labelled ribonucleotide triphosphates into an RNA probe. It is usual nowadays to use vectors that incorporate both S6 and T7 promotor regions in opposite orientation; in this way either sense or antisense RNA can be produced from the same construct, facilitating both positive and negative control RNA for the *in situ* hybridization experiment. Figure A3.3 illustrates the principle of this method of labelling.

Where long RNA probes are required more care is needed with the ribonucleoside triphosphate concentration (Melton *et al.* 1984). For use in *in situ* hybridization it is likely that shorter probes will be preferred. In common with the previous methods, the specific activity of the probe will depend upon the concentration and specific activity of the labelled triphosphates. Commercial products for the production of RNA probes are available.

Primer extension

Procedures to copy a single-stranded template in $5' \to 3'$ direction are widely used. An appropriate primer is hybridized to the template and this is extended enzymatically. Depending upon whether specific or random primers are used, either a specific or a generally labelled probe will result. The procedure can be employed with either linear or circular DNA, e.g. M13($+$) strand. The enzyme commonly used for this procedure is the *E. coli* polymerase I lacking the $5' \to 3'$ exonuclease activity; reverse transcriptase is also used for second strand synthesis. Depending upon the specific activity and concentration of radioactively labelled nucleotides, a wide range of probe-specific activities can be obtained including very high activities up to 4×10^9 dpm ^{32}P/μg of DNA. It has been shown that this method of radioactive probe preparation is not so sensitive to the purity of the initial DNA. (Feinberg and Vogelstein, 1984). A variety of commercial kits for a range of primer extension procedures are available.

End labelling of nucleic acids

End labelling techniques of nucleic acids were among the first methods used to obtain labelled probes. A number of methods are available.

T4 polynucleotide kinase will label the 5' OH of DNA or RNA. ATP gamma labelled is most frequently used as the donor, e.g. Maxim and Gilbert sequencing.

Terminal transferase catalyses the addition of deoxyribonucleotides to the 3' end of single or double stranded DNA. Either a single label can be

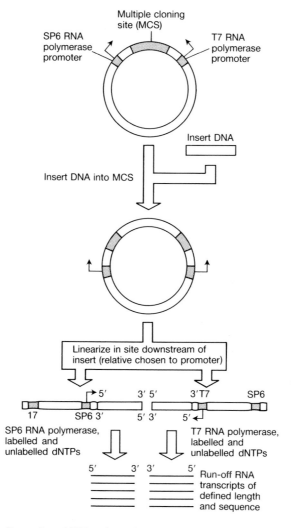

Preparation of RNA probes using a plasmid containing SP6 and
T7 polymerase promoters

Fig. A3.3.

added via a labelled cordycepin triphosphate or dideoxynucleotide, or
several labelled nucleotides can be added. A variety of ends, recessed or
blunt, can be labelled this way, however, it may be necessary to vary the
reaction conditions to suit the particular case in question.

A number of enzymes, Klenow or T4 polymerase can be used for end
repairing, e.g. to fill in 5′ overhangs which may be generated enzymatically.

Although most end labelling procedures are relatively straightforward, they are limited in the extent of radioactive labelling and hence specific activity of nucleic acid. Commercial kits for this are available. Table A3.3 indicates the range of ^{32}P-labelled activities that can be generated and how this is affected by the nature of the ends.

Table A3.3. Comparison of 3′-end labelling efficiencies with different labels

Type of 3′-end	Type of lambda digest	Counts incorporated per 10 picomoles of DNA ends (dpm $\times 10^{-6}$)		
		5′-ends labelled with [γ-^{32}P]ATP	3′-ends labelled with [α-^{32}P]ddATP	[α-^{32}P]KTP
Protruding	*Pst*I	1.9	16.6	6.0
Recessed	*Hind*III	24.5	12.6	3.5
Blunt	*Alu*I	11.4	61.5	16.5

Safety precautions for handling radioisotopes for *in situ* hybridization

Increasing emphasis is being placed today on radiological safety practice. Most of the isotopes used in *in situ* hybridization are β-emitters and their characteristics are given in Table 3.1.

As Table 3.2 shows, β particles can be stopped by the appropriate thickness of tissue dependent on the energy of the β particle. It is important to remember that if all of the energy of the β particle is stopped in tissue, very high local doses may result. In general, there are two major safety precautions that need to be observed with β-emitters, one is to avoid direct interaction with β particles themselves, e.g. a thickness of 0.1 mm of water is sufficient to stop ^{35}S β whereas more than 1 cm is required to attenuate ^{32}P β. The second major precaution is to avoid aerosol forming operations. Where there is *any* likelihood of aerosols such operations should be carried out in confined ventillated enclosures.

It is assumed that the customary precautions for handling isotopes will be followed, viz. that the laboratories are appropriately licenced; designated areas have been allocated; suitable monitoring equipment is available; and that proper radiological training and support is available.

A special video is available from Amersham International on the safe handling of radioactive materials which deals with all the major radioactive operations, e.g. vial opening, solution transfer, preparation of the work area, and where necessary, decontamination.

Because of the low activity levels generally required for *in situ* hybridization work, few additional special precautions are required. One further precaution should be the close proximity of a sink, or metal container, for radioactive liquid disposal; this allows ready disposal of hybridization buffer, stringency washes etc. Furthermore, any vessels used for hybridization and washing should be easily decontaminated for re-use. In addition, care should be taken at the hybridization stage to localize the radioactive solution around the area of the cells or tissue on the slide to minimize contamination of the buffer solution and vessel during hybridization.

The use of gloves and finger-tip dosimeters is strongly recommended for this work together with the use of appropriately designed forceps and similar devices to reduce operator exposure.

Procedures for the preparation of glass slides in *in situ* hybridization and auto-radiography

Glass microscope slides for *in situ* hybridization should be thoroughly cleaned before use by immersion in chromic acid for approximately 10 min and thoroughly washed to remove all traces of acid. Normally slides are then either siliconized via immersion in 5% dichloro-methyl silane in chloroform or coated with poly-L-lysine via a fresh 1 mg/ml solution. Where slides are siliconized they should be subsequently baked at approximately 250 °C for 4 h to remove RNases; when poly-L-lysine is used the baking should be carried out prior to coating. Once the slides have been heated, normal sterile handling techniques are required for all subsequent handling of slides, cells or tissues, fixation of samples on the slides, and subsequent hybridization and autoradiography.

Once the hybridization is complete, tissues or cells are dehydrated in graded ethanols, air-dried and prepared for autoradiography. Normal photographic precautions are required when handling films or emulsions in open conditions. Where stripping film is used it is coated onto the slide. Nuclear track emulsion usually requires further dilution according to the manufacturer's instructions—a pre-diluted ready-to-use form is now available from Amersham International. Where dilution is necessary, it is recommended that 0.6 M ammonium acetate is used rather than water to retain hybrid stability when working at high stringency. Particular care is needed in the preparation of the emulsion and dipping of the slides if an even, intimate, bubble-free coating is to be obtained and this is of crucial importance if cracks and defects are not to be introduced at this late stage in the assay.

The following protocol illustrates the important procedures that need to be followed to obtain evenly coated slides:—

---

1. Set the water bath to 43 °C and check the temperature.

2. Place the 50 ml beaker (containing a glass rod) and dipping vessel in the water bath to equilibrate. Keep the dipping vessel in the water bath throughout the procedure.

3. Arrange all the equipment in the check-list in an accessible manner.

4. Extinguish room light.

5. Switch on safelight.

6. Open emulsion stock and, using a plastic spatula, spoon approximately 15 ml into the 50 ml beaker.

7. Stir the gelled emulsion gently in the beaker using the glass rod. Within 10 min the emulsion should be completely melted. Alternatively, if all of the emulsion is to be used, the emulsion may be melted on its original container.

8. Pour the molten homogenous emulsion gently down the side of the dipping vessel until the vessel almost overflows.

9. Gently mix the emulsion in the dipping vessel with the glass rod and allow any bubbles to rise to the surface of the emulsion (about 5 min).

10. Remove the surface bubbles from the dipping vessel by displacing them. Push the rod to the bottom of the dipping vessel and the emulsion will overflow carrying with it most of the surface bubbles. If this procedure is not possible because of the volume of emulsion used or to avoid contaminating the waterbath, repeated dipping of blank slides will remove surface bubbles.

11. Quickly dip a blank subbed slide in the emulsion. Allow to drain briefly and hold up to the safelight and examine the emulsion layer for bubbles. If there are bubbles wait a further 10 min then repeat step 10 and check again. If emulsion coat is free of bubbles proceed to step 12.

12. Hold a test slide between the thumb and forefinger and dip it vertically into the vessel so as to cover $\frac{3}{4}$ of the slide area.

13. Hold the slide vertically for 5 s and then withdraw the slide from the emulsion at a steady rate keeping the slide vertical.

14. Pause at the edge of the dipping vessel and touch the slide to the rim of the vessel to avoid dripping the emulsion.

15. Still holding the slide vertically, drain the excess emulsion from the slide by standing for 5 s on a pad of tissues.

16. With a tissue, wipe the back of the slide free of emulsion taking care not to wipe the edges of the slide.

17. Place the slide emulsion-side uppermost on the chilled metal plate (wipe away any condensation on the plate first).

18. Leave the slides to gel on the plate for about 10 min.

19. Dip all the other slides in exactly the same manner.

20. Dry the slides in a rack in the dark at room temperature with a relative humidity of 60 per cent for 2–3 hours. Driers sold expressly for this purpose may be used—drying in this case takes approximately 30 min.

21. Remove the slides from the dryer and place them in the plastic slide box together with a small tissue bag of activated silica gel.

22. Seal the box with black insulating tape and place the slide box inside another box with an identifying label.

23. Seal up emulsion and return to container.

24. Switch on room lights.

25. Place slide box in a refrigerator at 4 °C for exposure.

References

Feinberg, A. P. and Volgelstein, B. (1984). A technique for radiolabelling DNA restriction endonuclease fragments to high specific activity. Addendum *Analytical Biochemistry*, **132**, 166–7.

Melton, D. A., Krieg, P. A., Rebagliati, M. R., Maniatis, T., and Zinn, K. (1984). Efficient in-vitro synthesis of biologically active RNA and RNA hybridization probes from plasmids containing a bacteriophage SP6 promoter. *Nucleic Acids Res.*, **12**, 7035–56.

Rigby, P. W. J., Diekmann, M., Rhodes, C., and Berg, P. (1977). Labelling deoxyribonucleic acid to high specific activity in vitro by nick translation with DNA polymerase I. *Journal of Molecular Biology*, **113**, 237–51.

4

Non-radioactive probes: preparation, characterization, and detection

V. T-W. CHAN AND J. O'D. McGEE

The need for developing non-radioactive hybridization techniques has been realized for more than 10 years simply because of the intrinsic disadvantages of radioisotopes (such as personnel and disposal problems). Various methods have been described for preparing non-radioactive hybridization probes, including chemically modified DNA molecules and cross-linking proteins to DNA molecules; the general utility of most of them has still to be demonstrated. Of the non-radioactive methods, biotinylated probes show promising results in various applications, e.g. filter (Chan *et al.* 1985) and *in situ* hybridization (Burns *et al.* 1985). Therefore, this chapter will be confined to the use of biotinylated probes. Digoxigenin-labelled probes, which are equally good for non-isotopic *in situ* hybridization (NISH) are discussed in Chapter 12.

Preparation of biotinylated probes

Hybridization probes can be divided into two classes, based on their chemical properties, i.e. DNA probes and RNA probes. Usually they are prepared from unlabelled nucleic acid molecules either enzymatically or chemically. There are several different methods available in both categories. The choice of a suitable one very much depends on the purpose and the requirement of a particular experiment. In general, enzymatic methods are relatively more satisfactory, whereas chemical methods are simpler and easier to perform. In enzymatic methods, a biotinylated analogue of a nucleotide is used to substitute for its native counterpart in the labelling reactions. For preparing DNA probes, either biotinylated dUTP or dATP can be used, whilst biotinylated rUTP is used in the preparation of RNA probes. In chemical methods, biotin is usually linked to highly reactive molecules. Under certain conditions, these biotinylated molecules react with one of the functional groups of nucleic acids (e.g. amine group). The suitability of biotinylated nucleic acid molecules as hybridization probes depends on two parameters: degree of label incorporated and the length of labelled fragments. Hybrids of biotinylated probes have a lower T_m than their native counterparts. The more biotin residues incorporated, the greater the decrease in the T_m (see below). There are different ways of controlling these two parameters in various labelling reactions. The principles as well as the characteristics of the end products of these labelling reactions will be described below. Based on this information and the specific requirement of the experiment, one can rationalize the choice of a suitable method.

Enzymatic methods

Nick translation

Nick translation (Rigby *et al.* 1977) is probably the most commonly used method for labelling hybridization probes. The designation is sometimes confused with the translation of proteins. In fact, it is not related to the translation of protein from mRNA. Two enzymes, namely, bovine pancreatic deoxyribonuclease I (DNase I) and *E. coli* DNA polymerase I (Pol I) are used in this reaction. DNase I is an endonuclease that hydrolyses double-stranded or single-stranded DNA to a complex mixture of mono- to oligonucleotides with 5′-phosphate termini. In the presence of Mg^{2+}, DNase I attacks each strand of DNA independently, resulting in single-stranded nicks. The sites of cleavage are distributed in a statistically random fashion. The degree of hydrolysis (or nicking) can be controlled by the concentration of DNase I in the reaction. The holoenzyme of Pol I has three different enzymatic activities: (a) $5′ \rightarrow 3′$ polymerase, (b) $5′ \rightarrow 3′$ exonuclease, (c) $3′ \rightarrow 5′$ exonuclease. When there is single-stranded nick in the DNA molecule, the exonuclease activity of Pol I hydrolyses the fragment 3′ to this nick in $5′ \rightarrow 3′$ direction. At the same time its polymerase activity sequentially incorporates nucleotide residues in the same direction to the 3′ hydroxy terminus of the fragment 5′ to the nick exactly complementary to the template strand (in this way, it is used as primer) in the presence of all four deoxyribonucleoside triphosphates.

The results of these two reactions is the movement of the nick in the $5′ \rightarrow 3′$ direction along the DNA strand, whose nucleotide sequence is basically unchanged. This movement of the nick along the DNA strand is the origin of the name nick translation. If one of the deoxyribonucleoside triphosphates is labelled (in this case, either bio-dUTP or bio-dATP), the reaction products are labelled DNA molecules, resulted from the replacement of original nucleotide residues by labelled nucleotides, with single-stranded nicks. As mentioned above, the frequency of the nicks in the DNA molecules can be controlled by the concentration of DNase I in the reaction mixture.

After denaturation of the labelled DNA molecules, single-stranded fragments with different lengths and sequences are obtained. The degree of label incorporated can be controlled by the concentration of labelled nucleotide, the concentration of Pol I, and the length of reaction time. Since biotinylated analogues of nucleotide are poorer substrates for Pol I than their native counterparts, its concentration is generally used at 20 μM which is ten times the minimum concentration of nucleotides for Pol I reaction. Since DNase I introduces nicks in both strands of DNA molecules in a virtually random fashion, theoretically the probe is uniformly and completely labelled. In fact, if the reaction is optimally controlled, nick-translated probes give the highest sensitivity compared to probes labelled with other methods. However, both the purity of DNA samples and the quality of nick translation reagents are very critical for achieving optimally nick-translated probes, which should have the following characteristics: (a) after denaturation, the length of single-stranded fragments should be within a small range, and (b) all the fragments should be uniformly and optimally labelled (see below). Practically, however, it is not easy to reproduce the result of nick translation simply, because it depends on the combination of the reactions of two enzymes that may be selectively inhibited by different contaminants in the DNA samples. Furthermore, the contaminants and their concentrations may vary in different DNA preparations. Therefore, for a particular DNA sample, the conditions of nick translation may have to be optimized. The parameters are the concentration of DNase I and Pol I. In general, different concentrations of DNase I and Pol I are used in various reactions, which are allowed to proceed for a fixed period of time. The combination of enzyme concentrations that gives optimally labelled probes is then used for the subsequent labelling reaction. Since biotinylated DNA probes can be stored for long periods (at least 2 years), large-scale labelling reactions can also be performed and the labelled probe is stored under optimal conditions (e.g. -70 °C). In this way, the same batch of labelled probe is used in all hybridization reactions and thus the results are more reproducible and reliable. However, optimization of reactions for many DNA samples can be very time-consuming. Therefore, use of DNA samples and nick translation reagents of high quality is strongly recommended.

Under well controlled conditions, this method can be used to label virtually any double-stranded DNA molecule and the resulting probes usually give highly satisfactory results.

Random-primed labelling

Random-primed labelling was developed in 1983 (Feinberg and Vogelstein 1983) and has gradually gained popularity since then. By contrast with nick translation, only the Klenow fragment of *E. coli* DNA polymerase I (Pol I) is used and no DNase I is needed in the reaction. The Klenow fragment differs from the holoenzyme of Pol I in that it lacks the $5' \rightarrow 3'$ exonuclease activity. In a typical random-primed labelling reaction, the DNA sample is first heat-denatured and the DNA can then be used as a template in the subsequent synthesis reaction. Hexadeoxyribonucleotides, which have no sequence specificity, all four deoxyribonucleoside triphosphates, and the Klenow fragment of Pol I are added to initiate the synthesis reaction. The initial event in the synthesis reaction is the binding of hexanucleotides to the denatured DNA strands (template). In this way, these short oligo-nucleotides function as primers for DNA synthesis. Since these short oligo-nucleotides have no sequence specificity, they bind to the template in a random fashion. In the presence of all four deoxyribonucleoside triphos-phates, the Klenow fragment incorporates nucleotide residues to the primers in $5' \rightarrow 3'$ direction complementary to the template strands. If a labelled deoxyribonucleoside triphosphate (in this case, bio-dUTP or bio-dATP) is used, the newly synthesized strands will become labelled by the incorporation of the labelled nucleotides. Since the synthesis of the com-plementary strands is primed at random sites of the template strands, this reaction is called random-primed labelling. In this synthesis reaction, both strands of the DNA molecules are used as templates; therefore, the labelled strands consist of species of two complementary strands.

The length of labelled strands can be controlled by varying the ratio of primer to template. High concentration of primer gives short labelled strands and vice versa. However, the ratio of primer to template producing optimal length of labelled probe has to be determined experimentally because the distribution of different nucleotides in a given DNA fragment is not equal, i.e. there may be A-T or G-C rich stretches in the DNA sequences. The reaction can be almost driven to completion by prolonging the reaction time because the concentration of nucleotides in the reaction mixture is generally in large excess. It should be realized that even when the reaction is driven to completion the final product of the reaction consists of labelled (synthesized) and unlabelled template strands. The unlabelled strands inevitably compete with the labelled strands in hybridization reactions. Assuming that the networking effect (see Chapter 1) is exactly the same as with nick-translated probes, then (unlike that of nick-translated probe) the complex of random-primed labelled probes is only partially labelled. This is probably the reason why optimally nick-translated probes have higher sensi-tivity than random-primed labelled probes. However, compared to nick translation, this method is more reliable and reproducible because the Klenow fragment is more resistant to inhibition by contaminants in DNA samples. In fact, it works reasonably well even in the presence of agarose, which is a potent inhibitor of many enzymes used in molecular biology.

This method can be used to label any linear DNA molecule. Supercoiled DNA molecules have to be linearized before labelling because this form of DNA molecule re-anneals extremely fast, resulting in a rapid decrease in the effective concentration of template strands. This labelling method is particularly useful for DNA samples extracted from agarose gels especially short DNA fragments, which usually gives poor results with nick-trans-lation.

In vitro transcription

Unlike the previous two methods, this method produces single-stranded RNA molecules with identical sequence to the cloned DNA fragment. However, in this method, the DNA fragment of interest has to be cloned in vectors containing phage transcription promoters that can initiate the transcription of the DNA fragment in the presence of the corresponding RNA polymerase. Different phage transcription promoters can be used for this purpose. However, three of them, T3, T7 and SP6 are most commonly used. In some cases, two different phage transcription promoters are placed on both sides of the polylinker cloning sites of the vectors. These two transcription promoters are arranged in opposite orientation so they can be used to initiate the transcription of the coding (sense) and non-coding strands (anti-sense) of the DNA insert respectively (see Chapter 6).

The recombinant plasmid can be prepared with standard methods. In the presence of all four ribonucleoside triphosphates and the corresponding phage RNA polymerase, the RNA transcripts are then synthesized. If a labelled ribonucleoside triphosphate (bio-rUTP) is included in the reaction mixture, the synthesized transcripts will become labelled. In general, after the reaction is complete, the DNA template is hydrolysed by DNase I while the labelled RNA is resistant to this treatment. The RNA can then be used as a hybridization probe. Since bio-rUTP is a poor substrate for SP6 RNA polymerase, this enzyme is not suitable for the synthesis of biotinylated RNA probes. On the other hand, RNA polymerase of T3 and T7 work well with bio-rUTP. An alternative way to use the SP6 system is the substitution of UTP by its allylamine derivative in the transcription reaction followed by the biotinylation of the synthesized RNA molecules containing abundant allylamine groups with biotin ester.

The size of the synthesized probe can be controlled by introducing a gap in the DNA template by restriction digestion, at different distances downstream of the transcription initiation site, at which the transcription is terminated. In this way, a different fraction of the insert is selectively transcribed. Alternatively, the size of full-length transcripts can be reduced to the desired length by controlled alkaline hydrolysis.

With this method, it is possible to synthesize highly labelled RNA probes. Furthermore, every RNA strand is labelled, thus eliminating competition between labelled and unlabelled molecules in hybridization reactions. The single-stranded nature of the transcribed RNA probes also prevents reannealing of the complementary strands, as occurs with double-stranded DNA probes. By using different transcription promoters and RNA polymerases, sense or antisense RNA transcripts can be selectively synthesized that are of proven usefulness in studies of gene expression and inhibition. Hybrids of target sequences and RNA probes have a higher T_m than that of DNA probes. Moreover, after hybridization, the non-specifically bound

RNA probes can be selectively removed by RNase treatment; hybridized RNA probes are resistant to RNase digestion. All these are advantages of this method that can increase both the sensitivity and specificity of the probes synthesized by this method.

End tailing

This method is based on the ability of terminal deoxyribonucleotidyl transferase (terminal transferase) to catalyse the addition of deoxyribonucleoside triphosphates to the 3′ hydroxy end of double- or single-stranded DNA in a template-independent fashion. Homopolymeric tails can thus be added to the DNA molecules. For instance, in the presence of bio-dUTP, long stretches of biotinylated poly(dU) are added to the 3′ termini of the DNA molecules whose internal sequences have not been replaced by any labelled nucleotides. The labelled DNA can then be used in hybridization reaction. The principle of 5′ labelling is similar (see Chapter 5).

Probes labelled by this method retain the original size (except for the additional tail). Furthermore, the hybrids resulting from the hybridization have virtually the same T_m as native DNA molecules. Owing to steric hinderance, however, only a small portion of the biotinyl residues on the homopolymeric tail is accessible to detection reagents, resulting in relatively lower sensitivity. The homopolymeric tail may also cross-hybridize to stretches of complementary sequences in the cellular genome or poly(A) tail of mRNA. These disadvantages limit the general utility of this method. However, it is useful for labelling short DNA fragments (100–200 bp) for which T_m is a critical factor. It has wide utility in characterizing polymerase chain reaction (PCR) products (see Chapter 12).

T4 DNA polymerase replacement

This method can be divided into two steps, namely, hydrolysis and resynthesis/repair. The repair step is optional. T4 DNA polymerase has two enzymatic activities: (a) 5′ → 3′ polymerase; and (b) 3′ → 5′ exonuclease. Its 3′ → 5′ exonuclease activity is approximately 200 times as active as *E. coli* DNA polymerase I and this property makes it suitable to be used in the replacement reaction. Like *E. coli* DNA polymerase I, the exonuclease activity of T4 DNA polymerase is inhibited by the presence of all four deoxyribonucleoside triphosphates. In this method, the DNA is first treated with T4 DNA polymerase in the absence of nucleotides resulting in the digestion of duplex DNA to produce molecules with recessed 3′ termini Deoxyribonucleoside triphosphates are subsequently added to initiate the resynthesis reaction. If a labelled nucleotide (e.g. bio-dUTP) is included in the reaction mixture, the DNA strands will be labelled as a result of the incorporation of the labelled nucleotides. After a period of time, a large quantity of unlabelled nucleotides may be added to drive the resynthesis reaction and the DNA is repaired to the original size. Under optimal conditions, the resynthesis reaction can proceed to near completion (i.e. the DNA molecules are almost completely repaired). The end product of these reactions is a population of DNA molecules that are fully labelled at their ends (or quite near the ends) but that contain progressively decreasing quantities of label toward their centre. After denaturation, every DNA strand consists of a labelled portion and an unlabelled one. If the repair step is included in the reaction, the original length of DNA fragments can be retained.

Probes can be highly labelled by the replacement reaction. However, it should be borne in mind that the probes are not uniformly labelled. The advantage of this method is that a specific region of a DNA strand can be selectively labelled which can be converted to a single-stranded hybridization probe by restriction digestion followed by purification from agarose gel.

Chemical methods

Photobiotin

The structure of photobiotin consists of a photoactive group, a spacer arm, and biotin (Forster *et al.* 1985). It is suggested that its photoactive group can react with a wide variety of functional groups but these groups are not well defined. Photobiotin can be used to biotinylate different biomolecules (e.g. DNA, RNA, and protein). Owing to its high reactivity, it reacts with a wide range of chemical molecules. In fact, in a typical photobiotinylation, the plastic tube used is also biotinylated. Photobiotin can efficiently label DNA and RNA in large quantity, but its non-specific background is higher than for enzymatically labelled probes. The problem may be partly due to the contaminants (which can easily be biotinylated) in the DNA samples. Another reason for background is the intrinsic structure of the molecule. Because of the cationic nature of its spacer arm, photobiotinylated molecules tend to bind to negatively charged molecules (e.g. nucleic acids) owing to electrostatic force. Even under high ionic strength, it still gives relatively high non-specific background. Furthermore, it is not as sensitive as enzymatically labelled probes. This is probably due to the low incorporation of biotinyl residues. Therefore, if a large quantity of probes is needed, and the overall sensitivity is not critical, photobiotin is an attractive alternative owing to its low cost.

Biotin hydrazide

Biotin hydrazide has been used to label proteins for a long time. However, its use in labelling DNA molecules was described only recently (Reisfed *et al.* 1987). It reacts with unpaired cytosine, catalysed by sodium bisulphite. Usually, it is necessary to incubate the reaction mixture for 24 h, which is considerably longer than required for any other method. However, the procedure *per se* is simple and easy. The sensitivity reported is comparable to the probes labelled by other methods. However, its sensitivity and reproducibility have to be confirmed. If its general utility for labelling hybridization probes is proved, it is certainly a very useful method.

Biotin ester

Biotin ester (ε-caproylamidobiotin-N-hydroxysuccinimide ester) can be used to label unpaired cytosine residues modified by transamination with sodium bisulphite and ethylenediamine. Detection of 1–2 pg of target sequence was reported (Viscidi *et al.* 1986). The procedure is simple and easy. However, as mentioned above, the sensitivity and reproducibility of this method have to be confirmed before its general utility in labelling hybridization probes can be proved. On the other hand, the incorporation of bio-rUTP into RNA by SP6 RNA polymerase is very low, whilst the enzyme can utilize allylamine UTP efficiently. In this *in vitro* transcription system, allylamine UTP is usually included in the reaction mixture. The modified RNA synthesized can then be biotinylated with biotin ester. Similarly, allylamine nucleotides can also be used in other enzymatic reactions (such as nick translation, etc.) and the products of these reactions are then biotinylated with biotin ester. Since allylamine nucleotides are better substrates than bio-nucleotides for many enzymes, and since biotinylation with biotin ester is more controllable, probes labelled in this way are more reliable and reproducible. However, it is a two-step procedure and thus more time-consuming.

Purification of labelled probes

After the labelling reaction, labelled probes are usually purified before being used in hybridizations. Three different methods can be used with satisfaction. Spin columns are a simple method in that the unincorporated labelled nucleotides are trapped in the matrix (e.g. Sephadex G-50) and the labelled probes, which are excluded from the pores of the matrix, are spun through the column and collected in the eluate. Conventional gel filtration columns (e.g. G-50) can also be used to purify labelled probes; however, in this method, fractions of eluate have to be collected. Thereafter, a small aliquot of each fraction is immobilized on a solid matrix (e.g. nitrocellulose) and tested for the presence of biotin residues. In general, conventional column chromatography is slightly better than spin columns but the former is more tedious.

Unincorporated labelled nucleotides can also be eliminated by ethanol precipitation in the presence of carrier DNA, which coprecipitates with the labelled probe whilst mononucleotides remain in the supernatant. Of these three methods, ethanol precipitation combines both simplicity and efficiency. Therefore, in our laboratories it is the method of choice for purifying labelled probes.

Characterization of labelled probes

The sensitivity of a biotinylated probe depends on two parameters, degree of substitution and size of labelled fragments.

Degree of substitution

Degree of substitution indicates the proportion of a species of native nucleotide residues substituted by labelled analogues. This can be determined by adding small amounts of tritiated nucleotide in the reaction mixture or by measuring the amount of biotinyl residues chemically. The sensitivity of detection correlates with incorporation of labels up to approximately 60 per cent substitution and then levels off. On the other hand, the melting temperature, T_m, as well as the hybridization efficiency inversely correlate with the degree of substitution. Therefore, the optimal sensitivity is a compromise between detection sensitivity and hybridization efficiency. In general, probes with 25–32 per cent substitution give highest sensitivity. To achieve this, the degree of substitution can be fine tuned by adding a small amount of the corresponding unlabelled nucleotide (e.g. dTTP for bio-dUTP) in the labelling reaction (Chan *et al.* 1985).

Size of labelled fragments

The size of single-stranded labelled fragments plays an important role in hybridization efficiency. Fragment size can be determined by fractionating labelled probes on a denaturing gel (e.g. glyoxal gel or alkaline gel) with molecular weight markers, followed by blotting and detection of the labelled fragments. In single-phase hybridization, fragments of approximately 800 nucleotides in length have highest efficiency. However, for *in situ* hybridization, small fragments (e.g. 200–300 nucleotides) are beneficial because of their higher penetration rate into the tissue section and thus give stronger signals.

Detection of biotinylated probes

Antibodies

Standard antibody procedures (e.g. indirect) can be employed for the detection of biotinyl residues in labelled nucleic acids. Both monoclonal and polyclonal antibiotin antibodies are commercially available (e.g. Dako, UK). There is no significant difference in sensitivity between them. However, antibody procedures are quite time-consuming and until recently their sensitivity was very limited (Herrington *et al.* 1989*a,b*; Chapter 12). The signal from the antibody procedure can be intensified approximately 100-fold with silver solution when peroxidase and DAB are used (Gallyos *et al.* 1982). After intensification, the sensitivity is adequate for detecting low copy numbers of target sequences on filters and tissue. However, it is well known that metallic silver also precipitates on proteins and nucleic acids, which are thus stained. Therefore, the conditions under which the specific signal is selectively intensified has to be determined. Furthermore, critical and careful interpretation of the results is needed owing to the possibility of non-specific staining.

Avidin/streptavidin

Avidin and streptavidin are biotin-binding proteins. The binding affinity of biotin and avidin/streptavidin is extremely high ($K_d = 10^{-15}$ M) which is almost equivalent to a covalent bond. Avidin is a glycoprotein present in egg white. Its relative molecular weight is 66 000 and it has a pI of 10.5. Under physiological conditions its carbohydrate moieties can bind to certain lectins, and owing to its positive charges, it also binds to negatively charged biomolecules (such as nucleic acids). These intrinsic properties significantly reduce the specificity of avidin as well as its general utility in detecting biotinylated probes. Contrarily, streptavidin (MWr = 60 000), which is a secreted protein of *Streptomyces avidinii*, does not have carbohydrate moieties and has a neutral pI. Therefore, it has much higher specificity than avidin. Thus, avidin has been almost completely superseded by streptavidin for the detection of biotinylated molecules, including biotinylated DNA.

Three different methods are commonly used in biotin/streptavidin detection systems: (1) streptavidin is used as a bridge between the primary target (biotinylated DNA) and the detectors (biotinylated enzymes, e.g. alkaline phosphatase); (2) a preformed complex of streptavidin and biotinylated enzyme, which is biotin-binding dominant, is used to detect the biotinyl residues of labelled nucleic acid; and (3) a conjugate that is prepared by chemically cross-linking enzymes to streptavidin is used for the detection. Theoretically, the second method should be the most sensitive, since a large complex of enzyme molecules is formed. However, all three methods have similar sensitivities, and in fact the first method is slightly better than the other two. It is, however, a two-step procedure and is thus more time-consuming than the one-step procedures of methods (2) and (3). On the other hand, the second method is less reliable and reproducible than the other two.

In general, alkaline phosphatase is the enzyme of choice. This can be detected by incubating with 5-bromo-4-chloro-3-indolyl phosphate (BCIP, enzyme substrate) and nitro blue tetrazolium (NBT, chromogen). So far, it has been demonstrated that the combination of the streptavidin–alkaline phosphatase–BCIP/NBT gives highest sensitivity in detecting biotinylated DNA. Changing any components of this combination invariably results in decreased sensitivity.

Perspectives

Oligonucleotide probes

Oligonucleotide probes are short DNA strands varying in size (usually less than 25 nucleotides). The optimal size for larger discrimination is 18–23 nucleotides long (see Chapter 5). Many different methods can be used to prepare biotinylated oligonucleotide probes (Murasugi and Wallace 1984; Kempe et al. 1985). This technique has several advantages: (1) it is easy to prepare labelled probes in large quantity; (2) it is possible to prepare probes for virtually any gene whose protein sequence is known; and (3) its specificity is so high that single point mutations can also be detected. They are widely used in various techniques such as library screening and blot hybridization. Oligonucleotide probes can also be used on in situ hybridization provided that the target sequences are highly abundant. This is because only two biotin molecules can be coupled to each oligomer, which then hybridize to the target sequences. In other words, there are only two biotin molecules for each copy of the target sequences. On the other hand, it is estimated that at least several hundred biotin molecules are needed to produce a visible signal on in situ hybridization. Therefore, this technique is not sensitive enough for detecting single-copy genes or low abundance transcripts.

Alternative labels and detection systems

Although biotin is probably the most sensitive non-isotopic reporter, other labels and detection methods have also been developed at about the same time (Tchen et al. 1984, Landegent et al. 1984). However, these systems are not substitutes for biotin. In fact, they are complementary to biotin in that several different labels and detection systems can be used on in situ hybridization at the same time, permitting simultaneous detection of several species of target sequences (see also Herrington et al. 1989a,b and Chapter 12). Theoretically, most haptens can virtually be used as non-radioactive labels for DNA and RNA probes. It is not necessary to prepare nucleotide analogues of these haptens. DNA and RNA can be modified by transamination or labelled with allylamine derivatives of the corresponding nucleotides. The modified DNA and RNA molecules can then be labelled with coupling reagents of these haptens (e.g. succinimidyl ester). In this way, DNA and RNA probes can be prepared by a simple two-step procedure. If different detection systems are used (e.g. antibodies for different haptens are labelled with different enzymes, or one is labelled with enzyme whilst the other is labelled with a fluorescent dye), it is possible to identify two (or more) species of target sequences in a single hybridization reaction. This technique has significant advantages especially in fine chromosomal localization of closely adjacent genes; differential expression of different or related genes during differentiation or embryonic development; and identification of closely related viruses in the same tissue section (see Chapter 12).

References

Burns, J., Chan, V. T.-W., Jonasson, J. A., Fleming, K. A., Taylor, S., and McGee, J. O'D. (1985). A sensitive method for visualizing biotinylated probes hybridized *in situ*: rapid sex determination on intact cells. *J. Clin. Pathol.*, **38**, 1085–92.

Chan, V. T.-W., Fleming, K. A., and McGee, J. O'D. (1985). Detection of sub-picogram quantities of specific DNA sequences on blot hybridization with biotinylated probes. *Nucl. Acids Res.*, **13**, 8083–91.

Feinberg, A. P. and Vogelstein, B. (1983). A technique for radiolabelling DNA restriction endonuclease fragments to high specific activity. *Analyt. Biochem.*, **132**, 6–13.

Forster, A. C., McInnes, J., Skingle, D. C., and Symons, R. H. (1985). Non-radioactive hybridization probes prepared by chemical labelling of DNA and RNA with a modified novel reagent, photobiotin. *Nucl. Acids Res.*, **13**, 745–61.

Gallyas, F., Gorce, T., and Merchenthaler, I. (1982). High-grade intensification of the end-product of the diaminobenzidine reaction for peroxidase histochemistry. *J. Histochem. Cytochem.*, **30**, 183–4.

Herrington, C. S., Burns, J., Graham, A. K., Evans, M., and McGee, J. O'D. (1989*a*). Interphase cytogenetics using biotin and digoxigenin-labelled probes I: relative sensitivity of both reporter molecules for HPV 16 detection in CasSki cells. *J. Clin. Pathol.*, **42**, 592–600.

Herrington, C. S., Burns, J., Bhatt, B., Graham, A. K., and McGee, J. O'D. (1989*b*). Interphase cytogenetics using biotin and digoxigenin-labelled probes II: simultaneous differential detection of human and HPV nucleic acids in individual nuclei. *J. Clin. Pathol.*, **42**, 601–6.

Kempe, T., Sundquist, W. I., Chow, F., and Hu, S. L. (1985). Chemical and enzymatic biotin-labelling of oligonucleotides. *Nucl. Acids Res.*, **13**, 45–57.

Landegent, J. E., Jansen in de Wal, N., Baan, R. A., Hoeijamkus, J. H. J., and van der Ploeg, M. (1984). 2-Acetylaminofluorene-modified probes for the indirect hybridocytochemical detection of specific nucleic acid sequences. *Exp. Cell. Res.*, **153**, 61–72.

Murasugi, A. and Wallace, R. B. (1984). Biotin-labelled oligonucleotides: enzymatic synthesis and use as hybridization probes. *DNA*, **3**, 269–77.

Reisfed, A., Rothenberg, J. M., Bayer, E. A., and Wilchek, M. (1987). Non-radioactive hybridization probes prepared by the reaction of biotin hydrazide with DNA. *Biochem. Biophys. Res. Commun.*, **142**, 519–26.

Rigby, P. W. J., Dieckmann, M., Rhodes, C., and Berg, P. (1977). Labelling deoxyribonucleic acid to high specific activity *in vitro* by nick translation with DNA polymerase I. *J. Mol. Biol.*, **113**, 237–51.

Tchen, P., Fuchs, R. P. P., Sage, E., and Leng, M. (1984). Chemically modified nucleic acids as immunodetectable probes in hybridization experiments. *Proc. Natl Acad. Sci. USA*, **81**, 3466–70.

Viscidi, R. P., Connelly, C. J., and Yolken, R. H. (1986). Novel chemical method for the preparation of nucleic acids for nonisotopic hybridization. *J. Clin. Microbiol.*, **23**, 311–17.

5

Oligonucleotide probes for *in situ* hybridization

R. LATHE

Introduction

Considerable advances in nucleotide chemistry now permit the routine synthesis and purification of oligonucleotides of defined sequence. Oligonucleotides make ideal hybridization probes because they can be designed to detect specific groups of genes, specific genes, or indeed specific alleles (genetic disease diagnosis) or serotypes (pathogen detection). However, the length of synthetic oligonucleotides (10–50 nucleotides) is far shorter than traditional probes based on large cloned DNA segments (0.5–5 kb) and this has important consequences for hybrid stability and probe specificity. The following presents a brief summary of major points to be considered in the design and utilization of oligonucleotide probes for *in situ* hybridization. For recent reviews of particular aspects, see Hames and Higgins (1985), Thein and Wallace (1986), Valentino *et al.* (1987), and Matthews and Kricka (1988).

Probe design

Length and uniqueness

A primary requirement of an oligonucleotide probe is that it must hybridize specifically to the target gene sequence and not to unrelated sequences. Statistically, shorter probes have an increased probability of finding perfect matches, by chance alone, in a given target genome. The K value for a given probe is defined as the number of matches expected in a random assortment of nucleotides of length L. When probe and target genome both have a 50 per cent G + C content:

$$K = (0.25)^l \times 2L$$

where l is the length of the probe. For a probe to be useful, the K value must not exceed 1, and in practice should be at most 0.1. Thus, if one considers that L for a mRNA population (or cDNA library) approximates to 10^7, and for a mammalian genome L approximates to 10^9, minimum

71

probe lengths can be calculated to be 14 and 17 nucleotides, respectively (Lathe, 1985).

However, hybridization experiments rarely discriminate between perfect matches (100 per cent homology) and near-perfect matches (95 per cent homology): by careful consideration of the equations for T_m and K given above, the K_b value (K value for the hybridization background) can be determined for a given probe. To confer hybridization selectivity, probe lengths must be increased to 16 and 20 nucleotides.

Probes from amino acid sequence data: genes are not random DNA

One of the tasks of molecular biology research is to identify genes encoding known proteins. Three approaches have been used to design oligonucleotide probes from amino-acid sequence data. In the first, probes containing mixtures of bases at each position of uncertainty (typically the codon third position) have been successfully used in a number of cases. In the second, consideration of codon preferences and other statistical rules permits the design of a unique 'optimal' (and rather longer) probe. Finally, single base-substituted probes containing deoxyinosine at each position of uncertainty have also been used.

In considering the second strategy it was determined that, from statistical analysis of known coding sequences, each position of uncertainty (the codon third position) could be predicted with a 46 per cent chance of success, and thus each triplet would correspond perfectly to the probe in 0.46 of cases. This result has an important consequence: there are only 20 amino acids, and the '*non-random*' K *value*, $K_{(nr)}$, can be determined as below:

$$K_{(nr)} = (0.46/20)^{1/3} \times 2L$$

which reduces to: $K_{(nr)} = (0.283)^l \times 2L$, a value rather greater than the basic $(0.25)^l \times 2L$ (see Fig. 5.1). Experimental test confirms this calculation (see Table 5.1).

Table 5.1.

Probe length	K values where $L = 10^7$ nucleotides			
	K	K_b	$K_{(nr)}$	$K_{b(nr)}$
12	1.2	9.4	5.3	62
14	0.07	1.0	0.43	7.2
16	0.0046	0.11	0.034	0.84
18	nd	0.011	0.0028	0.093
20	nd	0.0012	nd	0.01

K_b is the K value for the hybridization background; $K_{(nr)}$ is the K value in a collection of non-random (coding) sequences; $K_{b(nr)}$ is the non-random K_b value; nd, not determined.

(a)

$$\frac{1}{4} \quad \frac{1}{4} \quad \frac{1}{4} \quad \frac{1}{4} \quad \frac{1}{4} \quad \frac{1}{4} \quad \frac{1}{4} \quad \frac{1}{4} \quad \frac{1}{4} \qquad \left(\frac{1}{4}\right)^9 = \frac{1}{260\,000}$$

(b)

$$\frac{0.46}{20} \quad \frac{0.46}{20} \quad \frac{0.46}{20} \qquad \left(\frac{0.46}{20}\right)^3 = \frac{1}{82\,000}$$

(Mean codon third position certainty = 46 per cent)

Fig. 5.1. Diagrammatic presentation of the distinction between standard K-values and values for coding sequences for a 9-mer oligonucleotide. In (a) the K value is calculated by multiplying the individual nucleotide contributions, in (b) by multiplying the individual triplet contributions.

For this reason, when oligonucleotide probes are designed to detect gene sequences, probe length must be increased to confer specificity of hybridization, typically to 18 nucleotides for detecting particular mRNAs or cDNAs (Lathe 1985).

Finally, Taneja and Singer (1987) noted that similarly-sized oligonucleotides corresponding to different regions of a target mRNA gave *in situ* hybridization signals that greatly differed in intensity, suggesting that not all target sequences are equivalent.

Labelling of oligonucleotide hybridization probes

Oligonucleotides cannot take advantage of lattice-formation amplification (see Chapter 1) that occurs when a long double-stranded probe is employed in a hybridization experiment, and high-specific-activity labelling is necessary in order to develop a detectable hybridization signal. Traditionally, oligonucleotides may be labelled to high specific activity by:

(1) transfer of radioactive phosphate from $\gamma\text{-}^{32}\text{P}$-riboATP to the 5' end using polynucleotide kinase encoded by coliphage T4;

(2) internal labelling using a primer to copy part of the oligonucleotide into DNA using DNA polymerase (e.g. Klenow fragment of the *E. coli* DNA polymerase I) and $\alpha\text{-}^{32}\text{P}$-deoxytriphosphates, dNTPs substituted at the alpha position with an $\alpha\text{-}^{35}\text{S}$-thiophosphate group, or other radioactive dNTP derivatives (e.g. ^3H);

(3) template-independent polymerization involving tailing on to the 3' terminus using terminal transferase (from calf thymus) in the presence of α-labelled dNTPs; (note that probes carrying poly-T will hybridize to all mRNAs);

(4) labelling *in vitro* by iodination using Na^{125}I and a catalyst.

These procedures are summarized in Fig. 5.2.

In a comparison of 5'-end labelled and 3'-radioactively tailed oligo-nucleotide probes, Collins and Hunsaker (1985) were unable to detect significant differences in the thermal dissociation profiles. In addition, they showed that the hybridization sensitivity of probes tailed with α-thio-(^{35}S)-nucleotides (half-life of ^{35}S = 87 d) was comparable to that obtained with a short-lived 5'-(^{32}P)-labelled probe (half-life of ^{32}P = 14 d).

The need to avoid radioactivity in routine procedures is clear. Bases to which biotin has been coupled can be incorporated into oligonucleotides and take advantage of the avidin–biotin interaction to bind enzymes (e.g. peroxidase or alkaline phosphatase) with a convenient colour development reaction. A dual-specificity monoclonal antibody to biotin and to the enzyme may improve sensitivity and reduce background. A parallel development using digoxigenin-dUTP and enzyme-linked antibody to digoxigenin (Herrington *et al.* 1989*a,b*) appears to be equally sensitive. Poly-A tails added by terminal transferase could in principle also be detected by a coupled reaction of polynucleotide phosphorylase (to generate ADP), pyruvate kinase (to generate ATP), and luciferase (to generate detectable light) (see Gillam 1987). Finally, direct enzyme-linked oligonucleotides have been used in diagnosis of infective agents in clinical samples (Li *et al.* 1987; see also Urdea *et al.* 1987).

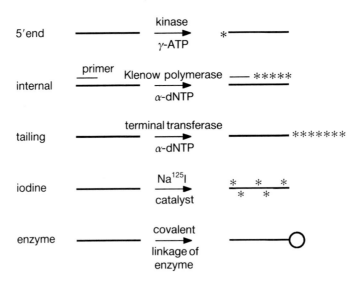

dNTP may be radioactively labelled in the α position or modified to permit non-radioactive detection

Fig. 5.2. Schematic presentation of the different methods of labelling oligonucleotide probes for *in situ* hybridization. *, radioactive label; O, enzyme linked to derivatized oligonucleotide.

One further technique is likely to prove of general application. Tecott *et al.* (1988) describe how an unlabelled oligonucleotide probe can be hybridized to mRNA present in a fixed tissue section. Subsequent incubation with reverse transcriptase and radiolabelled precursors permits the specific copying of the target mRNA species into complementary DNA. Rapid autoradiographic localization of proopiomelanocortin mRNA in rat pituitary was reported.

Probe hybridization

Calculating the melting temperature (T_m)

The temperature of a hybrid generally depends on the *salt* concentration and the base composition (Shildkraut and Lifson 1965). Increasing monovalent cation (e.g. Na^+) and increasing G-C content both stabilize the duplex formed. In addition, hybrid stability is decreased with decreasing duplex length or homology between the two strands (Thomas and Dancis 1973). Hybrid T_m can be estimated from the equation:

$$T_m(°C) = 16.6 \log M + 0.41(\%G + C) + 81.5 - 820/l - 1.2(100 - h)$$

where M is the monovalent cation concentration, $\%G + C$ the base composition, l the duplex length, and h the per cent interstrand homology (see Lathe 1985). In general, for probes with a G-C content of 50 per cent, and under salt conditions of $6 \times$ SSC, the equation reduces to:

$$T_m(°C) = 102 - 820/l - 1.2(100 - h)$$

For perfectly matched probes between 11 and 22 nucleotides in length, this equation can be approximated by:

$$T_m(°C) = 2 \text{ °C per A-T} + 4 \text{ °C per G-C}$$

(see Suggs *et al.* 1981). Values for $2 \times$ SSC are 8 °C lower than for $6 \times$ SSC.

In general, the inclusion of *formamide* in the hybridization medium reduces the melting temperature by approximately 0.7 °C per one per cent formamide. Note, however, that formamide destabilizes A/T rich hybrids disproportionately more than it does G/C-rich hybrids.

The above equations all deal specifically with DNA–DNA hybrids. It must be stressed that the thermal stability of RNA–DNA hybrids exceeds that of comparable DNA–DNA hybrids, particularly in the presence of formamide, and at 80% formamide the T_m of an RNA–DNA hybrid can exceed that of the cognate DNA–DNA hybrid by 20 °C. Finally, the thermal dissociation profile is strongly dependent upon the precise sequence of nucleotides in the probe. This is well illustrated by Buvoli *et al.* (1987), who report a difference of up to 7 °C in melting temperature between a 15-mer DNA–RNA duplex carrying a central T/A base pair and the same duplex possessing a central A/U base pair, the former being more stable. Albretsen *et al.* (1988) present experimental results obtained with a series of synthetic oligonucleotides ranging from 17 to 50 in length and report that the observed T_m often exceeds the calculated T_m by up to 10 °C, possibly because hybrids immobilized to a solid support may intrinsically possess a higher T_m. The same conclusion, for DNA–DNA hybrids formed *in situ*, has been arrived at by Herrington *et al.* 1990. For these reasons, the empirical formulae (above) are presented only as a rough guide to establishing the most appropriate hybridization conditions.

Hybridization conditions

In general, the maximum rate of hybridization takes place at T_m minus 10–25 °C (Bonner *et al.* 1973) although, with oligonucleotide probes, the rate of hybridization rarely presents an experimental problem. However, it remains standard practice to use long incubation times (e.g. overnight) and conditions approximating to T_m minus 20 °C. Importantly, the specificity of an oligonucleotide hybridization probe is often dictated not by the hybridization conditions but by the temperature, and particularly the duration, of stringent washing following hybridization (see below).

Allele specificity: oligonucleotide hybridization is kinetic

An advantage of oligonucleotide probes is that they may in principle hybridize to one gene type and not to a closely related gene type. Oligonucleotides can be designed to match perfectly with a mutant allele while hybrids with the wild-type allele will contain a destabilizing mismatch. Unfortunately, the situation is complicated by a number of parameters that must be taken into consideration.

First, longer probes give better hybridization selectivity (above). However, the longer a hybrid duplex the less a single mismatch will reduce the interstrand homology, and the less the T_m will be reduced. Thus, between the perfect and mismatched hybrids the T_m difference is less for longer probes (Table 5.2). In practice, a T_m difference of 5 °C is perhaps a reasonable detection limit, and probes exceeding 24 nucleotides are hence ruled out.

Table 5.2.

Probe length	K_b (genomic)	T_m (2×SSC)	T_m change 1 mismatch
14	7	35 °C	−9 °C
18	0.1	48 °C	−7 °C
22	nd	57 °C	−5.5 °C
26	nd	63 °C	−4.6 °C

nd, not determined.

Second, the destabilizing influence of a particular mismatch is dependent upon the nature of the mismatch, as well as on the particular flanking bases (and the orientation of the mismatch). Ikuta *et al.* (1987) report that G-T and G-A mismatches are far less effective in destabilizing a 19-mer duplex than A-A, T-T, C-T or C-A mismatches.

Third, oligonucleotide hybridization is not all-or-none and the kinetics of oligonucleotide hybridization must be examined. Hybridization is normally performed at a temperature 20–25 °C below the T_m of the hybrid because the hybridization rate is maximal under these conditions. This permits the formation of incorrect hybrids between the probe and fortuitous sequences sharing only around 80 per cent homology. Hybridization specificity is ensured by washing, for a limited period (e.g. 1 min) at a temperature close to the T_m of the desired hybrid (e.g. T_m minus 5 °C). Thus, the outcome of an oligonucleotide hybridization experiment depends more upon the probe–target dissociation rate than upon the calculated T_m. This can be determined from the results of Ikuta *et al.* (1987) (Table 5.3). Thus, wash times for a 10-fold difference in hybridization signal between perfect match and mismatched hybrid can be calculated (Table 5.4).

Therefore the calculation of T_m is used to determine the approximate temperature for optimal washing, but the wash time is equally, or more important.

Table 5.3. Observed dissociation half lives (min)

Temperature (°C)	Perfect match (A-T)	Mismatch (G-T)	Mismatch (A-A)
50	16.5	9.9	5.3
60	1.0	0.6	0.4

Standard T_m calculations taking G/C content into account give: perfect hybrid, T_m = approx. 61 °C; single mismatch, T_m = approx. 55 °C.

Table 5.4.

Temperature (°C)	Mismatch	Wash time for 10 × difference (min)	Per cent initial hybridization signal lost during washing
50	G-T	80	> 95
60	G-T	5	> 95
50	A-A	30	70
60	A-A	2	75

Oligonucleotides as probes

Many of the experimental approaches to probe design, labelling, and hybridization described above have been successfully used to detect particular target sequences in tissue sections (for instance, Arentzen *et al.* 1985; Uhl *et al.* 1985; Davis *et al.* 1986; Lewis *et al.* 1986*a,b*; Han *et al.* 1987; Standish *et al.* 1987; Largent *et al.* 1988; Zoeller *et al.* 1988). The recent development of oligonucleotide-directed enzymatic amplification of target nucleic acid sequences by the polymerase chain reaction (PCR) technique (see Saiki *et al.* 1988) opens a new approach to the detection of extremely rare sequences in tissue samples (e.g. Shibata *et al.* 1988). Finally, although allele-specific oligonucleotide probes have not been widely used in *in situ* hybridization experiments, their potential is demonstrated by the ability of specific oligonucleotides to detect genomic α-1-antitrypsin variants linked with emphysema; or HLA alleles linked to diabetes (e.g. Kidd *et al.* 1983; Todd *et al.* 1987; see also Alves *et al.* 1988).

In conclusion, the ease of synthesis and potential allele-specificity of oligonucleotides makes them agents of choice as hybridization probes. Difficulties in high-specific-activity labelling have so far restricted their use. However, the application of sensitive enzyme-linked non-radioactive assays to oligonucleotide probe technology is likely to lead to increased use of such probes both in the laboratory and in the clinic.

Acknowledgements

I would like to thank D. H. Wreschner for helpful comments on the manuscript.

References

Albretsen, C., Haukanes, B.-I., Aasland, R., and Kleppe, K. (1988). Optimal conditions for hybridization with oligonucleotides: a study with myc-oncogene DNA probes. *Analyt. Biochem.*, **170**, 193–202.

Alves, A. M., Holland, D., Edge, M. D., and Carr, F. J. (1988). Hybridization detection of single nucleotide changes with enzyme-labelled oligonucleotides. *Nucl. Acids Res.*, **16**, 8722.

Arentzen, R., Baldino, F., Davis, L. G., Higgins, G. A., Lin, Y., Manning, R. W., and Wolfson, B. (1985). *In situ* hybridization of putative somatostatin mRNA within hypothalamus of the rat using synthetic oligonucleotide probes. *J. Cell Biochem.*, **27**, 415–22.

Bonner, T. I., Brenner, D. J., Neufeld, B. R., and Britten, R. J. (1973). Reduction in the rate of DNA reassociation by sequence divergence. *J. Mol. Biol.*, **81**, 123–35.

Buvoli, M., Biamonti, G., Riva, S., and Morandi, C. (1987). Hybridization of oligonucleotide probes to RNA molecules: specificity and stability of duplexes. *Nucl. Acids Res.*, **15**, 9091.

Collins, M. L. and Hunsaker, W. R. (1985). Improved hybridization assays employing tailed oligonucleotide probes: a direct comparison with 5′-end labelled oligonucleotide probes and nick translated plasmid probes. *Analyt. Biochem.*, **151**, 211–24.

Davis, L. G., Arentzen, R., Reid, J. M., Manning, R. W., Wolfson, B., Lawrence, K. L., and Baldino, F. (1986). Glucocorticoid-sensitive vasopressin mRNA synthesis in the paraventricular nucleus of the rat. *Proc. Natl Acad. Sci. USA*, **83**, 1145–9.

Gillam, I. C. (1987). Non-radioactive probes for specific DNA sequences. *Trends Biotechnol.*, **5**, 332–4.

Hames, B. D. and Higgins, S. J. (eds.) (1985). *Nucleic acid hybridization, a practical approach*. IRL Press, Oxford.

Han, V. K. M., D'Ercole, A. J., and Lund, P. K. (1987). Cellular localization of somatomedin (insulin-like growth factor) messenger RNA in the human fetus. *Science*, **236**, 193–7.

Herrington, C. S., Burns, J., Graham, A. K., Bhatt, B., and McGee, J. O'D. (1989*a*). Interphase cytogenetics using biotin and digoxigenin labelled probes II: simultaneous detection of two nucleic acid species in individual nuclei. *J. Clin. Pathol.*, **42**, 592–600.

Herrington, C. S., Burns, J., Graham, A. K., Evans, M. F., and McGee, J. O'D. (1989*b*). Interphase cytogenetics using biotin and digoxigenin labelled probes I: relative sensitivity of both reporters for detection of HPV in Laski cells. *J. Clin. Pathol.*, **42**, 601–6.

Herrington, C. S., Burns, J., Graham, A. K., and McGee, J. O'D. (1990). Evaluation of discriminative stringency conditions for the distinction of closely homologous DNA sequences in clinical biopsies. *Histochem. J.* (In press.)

Ikuta, S., Takagi, K., Wallace, R. B., and Itakura, K. (1987). Dissociation kinetics of 19-base paired oligonucleotide-DNA duplexes containing different single mismatched base pairs. *Nucl. Acids Res.*, **15**, 797–811.

Kidd, V. J., Wallace, R. B., Itakura, K., and Woo, S. L. C. (1983). Alpha-1-antitrypsin deficiency detection by direct analysis of the mutation in the gene. *Nature*, **304**, 231–4.

Largent, B. L., Jones, D. T., Reed, R. R., Pearson, R. C. A., and Snyder, S. H. (1988). G protein mRNA mapped in rat brain by *in situ* hybridization. *Proc. Natl Acad. Sci. USA*, **85**, 2864–8.

Lathe, R. (1985). Synthetic oligonucleotide probes deduced from amino acid sequence data: theoretical and practical considerations. *J. Mol. Biol.*, **183**, 1–12.

Lewis, M. E., Sherman, T. G., Burke, S., Akil, H., Davis, L. G., Arenten, R., and Watson, S. J. (1986*a*). Detection of proopiomelanocortin mRNA by *in situ* hybridization with an oligonucleotide probe. *Proc. Natl Acad. Sci. USA*, **83**, 5419–23.

Lewis, M. E., Arentzen, R., and Baldino, F. (1986*b*). Rapid, high-resolution *in situ* hybridization histochemistry with radioiodinated synthetic oligonucleotides. *J. Neurosci. Res.*, **16**, 117–24.

Li, P., Medon, P. P., Skingle, D. C., Lanser, J. A., and Symons, R. A. (1987). Enzyme-linked synthetic oligonucleotide probes: non-radioactive detection of enterotoxigenic *Escherichia coli* in faecal specimens. *Nucl. Acids Res.*, **15**, 5275–87.

Matthews, J. A. and Kricka, L. J. (1988). Analytical strategies for the use of DNA probes. *Analyt. Biochem.*, **169**, 1–25.

Saiki, R. K., Gelfland, D. H., Stoffel, S., Scharf, S. J., Higuchi, R., Horn, G. T., Mullis, K. B., and Erlich, H. A. (1988). Primer-directed enzymatic amplification of

DNA with a thermostable DNA polymerase. *Science*, **239**, 487–91.

Schildkraut, C. and Lifson, S. (1965). Dependence of the melting temperature of DNA on salt concentration. *Biopolymers*, **3**, 195–208.

Shibata, D., Martin, W. J., and Arnheim, N. (1988). Analysis of DNA sequences in forty-year-old paraffin-embedded thin-tissue sections: a bridge between molecular biology and classical histology. *Cancer Res.*, **48**, 4564–6.

Standish, L. J., Adams, L. A., Vician, L., Clifton, D. K., and Steiner, R. A. (1987). Neuroanatomical localization of cells containing gonadotropin-releasing hormone messenger ribonucleic acid in the primate brain by *in situ* hybridization histochemistry. *Mol. Endocrinol.*, **1**, 371–6.

Suggs, S. V., Hirose, T., Miyake, T., Kawashima, E. H., Johnson, M. J., Itakura, K., and Wallace, R. B. (1981). Use of synthetic oligodeoxyribonucleotides for the isolation of specific cloned DNA sequences. *ICN-UCLA Symb. Mol. Cell. Biol.*, **23**, 682–93.

Taneja, K. and Singer, R. H. (1987). Use of oligodeoxynucleotide probes for quantitative *in situ* hybridization to actin mRNA. *Analyt. Biochem.*, **166**, 389–98.

Tecott, L. H., Barchas, J. D., and Eberwine, J. H. (1988). *In situ* transcription: specific synthesis of complementary DNA in fixed tissue sections. *Science*, **240**, 1661–4.

Thein, S. L. and Wallace, R. B. (1986). The use of synthetic oligonucleotides as specific hybridization probes in the diagnosis of genetic disorders. In *Human genetic diseases, a practical approach* (K. E. Davies, ed.), pp. 33–50. IRL Press, Oxford.

Thomas, C. A. and Dancis, B. M. (1973). Ring stability. *J. Mol. Biol.*, **77**, 43–55.

Todd, J. A., Bell, J. I., and McDevitt, H. O. (1987). HLA-DQb gene contributes to susceptibility and resistance to insulin-dependent diabetes mellitus. *Nature*, **329**, 601–4.

Uhl, G. R., Zingg, H. H., and Habener, J. F. (1985). Vasopressin mRNA *in situ* hybridization: localization and regulation studied with oligonucleotide cDNA probes in normal and Brattleboro rat hypothalamus. *Proc. Natl Acad. Sci. USA*, **82**, 5555–9.

Urdea, M. S., Running, J. A., Horn, T., Clyne, J., Ku, L., and Warner, B. D. (1987). A novel method for the rapid detection of specific nucleotide sequences in crude biological samples without blotting or radioactivity: application to the analysis of hepatitis B virus in human serum. *Gene*, **61**, 253–64.

Valentino, K. L., Eberwine, J. H., and Barchas, J. D. (eds) (1987). In situ *hybridization: applications to neurobiology*. Oxford University Press, Oxford.

Zoeller, R. T., Seeburg, P. H., and Young, W. S. (1988). *In situ* hybridization histochemistry for messenger ribonucleic acid (mRNA) encoding gonadotrophin-releasing hormone (GnRH): effect of estrogen on cellular levels of GnRH mRNA in female rat brain. *Endocrinology*, **122**, 2570–7.

6

Principles and applications of complementary RNA probes

S. J. GIBSON AND J. M. POLAK

The use of *in situ* hybridization for the detection of messenger (mRNA) molecules is increasing. Since cells manufacture a wide variety of proteins subserving complex functions (e.g. structural proteins, peptide neurotransmitters, hormones, enzymes, receptors, etc.), it is predictable that each cell will contain innumerable kinds of mRNA. Some mRNAs will be very abundant and present in most cells (e.g. those encoding housekeeping proteins) while others are likely to be relatively specific to certain cell types and comprise only a small fraction of the total cellular mRNA content. The exquisite specificity of complementary base pairing between a nucleic acid probe and its mRNA allows the detection *in situ* of such low-abundance mRNAs.

The different types of nucleic acid probe are discussed in several chapters in this volume. In this chapter discussion will centre on the use of RNA probes for *in situ* hybridization, their principles and applications.

The use of RNA probes for *in situ* hybridization has been pioneered largely by Angerer and colleagues (see Cox *et al.* 1984). Riboprobes are single-stranded RNA molecules produced from a cloned cDNA, that has been introduced into a specifically designed plasmid reverse-transcription system. Because these probes are single-stranded, unlike double-stranded DNA probes, they do not re-anneal in solution, so a greater percentage of probe is available for hybridization giving stronger signals than cDNA probes (see Cox *et al.* 1984). In addition, the cRNA–mRNA hybrids produced in solution are more stable than corresponding cDNA–mRNA hybrids (Wetmur *et al.* 1981), but this has not been proved for hybrids formed *in situ*. The labelled probes are of fixed length and high specific activity, owing to the way the transcription reaction takes place. However, these probes are sometimes 'stickier' than DNA probes and produce a higher degree of non-specific binding to tissue. This problem can be circumvented by the use of enzymes in the post-hybridization solution.

Recently, Wolf *et al.* (1987) have suggested the construction of riboprobes by inserting oligonucleotides of determined length in appropriate vectors to generate cRNA molecules. These 'oligoriboprobes' have now been successfully tried for *in situ* hybridization (Denny *et al.* 1988).

General protocol for *in situ* hybridization using complementary RNA probes

Although protocols are attached in the appendix, the list below is included to delineate the main steps of the procedure. The various steps will be discussed in sequence.

1. Tissue Treatment
    (i)   Fixatives
    (ii) Tissue permeabilization after fixation
2. Construction of cRNA probes
3. Probe labelling
4. Background reduction treatment
5. Prehybridization
6. Hybridization
    (i)   Probe concentration
    (ii) Probe length
    (iii) Stringency
    (iv) Stability of hybrids
    (v)   Kinetics
7. Post-hybridization treatments
8. Detection system
    (i)   Isotopic probes (X-ray film-emulsion dipping)
    (ii) Non-isotopic (enhancing detection systems)
9. Determination of specificity
10. Quantification

1. Tissue treatment

(i) Fixatives

The aims of fixation for *in situ* hybridization include avoidance of nucleic acid loss from tissues, preservation of morphology, and enhancement of probe penetration. This loss is frequently at odds with the other two aims, since sometimes tissue permeabilization may induce leaching of RNA and loss of tissue morphology. Fixation can be carried out by perfusion or immersion, or after fresh frozen sections are obtained. Paraffin-embedded fixed tissues have been employed successfully. Precipitating fixatives such as ethanol/acetic acid or Carnoy's fluid and cross-linking fixatives such as paraformaldehyde, formaldehyde, and glutaraldehyde have all been used. Precipitating fixatives provide the best probe penetration, but may permit the loss of RNA and provide imperfect tissue morphology (Lawrence and Singer 1985). The aldehyde fixatives promote cross-linking of proteins and

generally provide better RNA retention and morphology than the precipitating fixatives. Glutaraldehyde provides the best RNA retention and tissue
morphology, but because of extensive cross-linking, probe penetration is a
problem. Fixation in 4% paraformaldehyde, by perfusion or immersion of
tissue or of fresh cryostat sections have proved to be a popular compromise
between permeability and RNA retention for the demonstration of mRNA
species using cRNA probes.

(ii) Tissue permeabilization after fixation

Permeabilization is necessary in order to enhance accessibility of target
mRNAs to the probe, and the need for it varies according to the type and
extent of fixation, the tissue used, the section thickness and the length of the
probe. For instance, tissue heavily cross-linked (e.g. after glutaraldehyde
fixation) will require extensive permeabilization. This treatment may
involve exposure to dilute acids, detergents, alcohols, and protease such as
proteinase K, pronase, or pepsin. Excessive deproteinization, however, may
result in decreased RNA retention and deterioration of morphology.

2. Construction of cRNA probes

RNA probes are obtained by inserting a specific cDNA sequence into an
appropriate transcription vector, containing an RNA polymerase promoter.
Vectors of this type include the plasmids pSP64 and pSP65, the Gemini
vectors and Bluescript. The last two possess two different promoters
located on either side of the multiple cloning site. The plasmids pSP64 and
pSP65 differ in the orientation of the multiple cloning site in relation to the
SP6 promoter.

Transcription of 'sense' (control) and 'antisense' (cRNA) strands can thus
be obtained, either by inserting the specific cDNAs into two separate
plasmids (pSP64 and pSP65) or into those containing two different
promoters on either side.

For transcription, the circular plasmid is first linearized by a restriction
endonuclease digestion and under appropriate reaction conditions the
polymerase will incorporate free ribonucleotides into single-stranded
cRNA transcripts of the DNA insert (see Fig. 6.1 and protocol).

3. Probe labelling

Labelling is carried out using isotopes or non-isotopic substances. The
most commonly used isotopes include ^{32}P, ^{35}S, and ^{3}H. The most commonly
used non-isotopic label is biotin (vitamin H of B_2 complex) which will bind
to the glycoprotein avidin with very high affinity (see Chapter 4). Non-
isotopic labels offer several advantages, including speed, safety, and high
resolution. Non-isotopic labelling is carried out by the use of a biotinylated
uridine 5′-triphosphate (rUTP) derivative containing a carbon spacer arm,

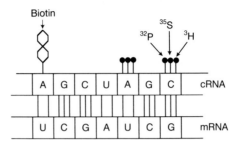

Fig. 6.1. Diagram to illustrate various methods available for labelling complementary RNA probes for the technique of *in situ* hybridization following linearization of the cDNA.

which increases availability of the biotin. Incorporation is achieved by the classical nick translation method or in vitro transcription (Chapter 4). Reaction product detection is achieved by the use of avidin or antibiotin sera.

4. Background reduction treatment

Background may be due to a number of factors including formation of imperfect duplexes with non-homologous nucleic acids, electrostatic interactions among charged groups, physical entrapment of probe in the three-dimensional lattice of the tissue section, and artefacts of the detection system. The non-specific retention of probe in tissue sections may be reduced by methods known to decrease tissue 'stickiness' or by blocking non-specific binding of probe. Posthybridization enzyme treatments and washes (discussed later) also help to reduce background. The 'stickiness' of tissue may depend in part on electrostatic attraction between the hybridization probe and basic proteins in the tissue section. This possible source of background is sometimes reduced by the use of 0.25% acetic anhydride, which blocks basic groups by acetylation (Tecott *et al.* 1987). Some investigators also acetylate slides and coverslips to minimize non-specific adherence of probes to glass.

5. Prehybridization

A prehybridization step is also designed to decrease background labelling. The components of the prehybridization mix are intended to saturate sites in the tissue section that might otherwise bind to nucleic acid unspecifically. These include ficoll, bovine serum albumin, and polyvinyl pyrrolidone. Sodium pyrophosphate, nucleic acids, and ethylenediamine tetra-acetic acid (EDTA) are also added to decrease non-specific nuclei acid interactions.

6. Hybridization

Factors to be considered include probe concentration, length, stringency of hybridization conditions, stability of hybrids, kinetics of *in situ* hybridization and preservation of tissue morphology (Tecott *et al.* 1987). Hybridization mixtures generally include all the components of prehybridization mixtures, as well as labelled probe and dextran sulphate. Dextran sulphate can also greatly amplify hybridization signals by a mechanism thought to be due to exclusion of DNA from the volume occupied by the polymer, thus increasing the effective concentration of the probe. 'Networking' is probably not operative with dextran sulphate and cRNA probes, as it is with nick translated plasmids (see Chapters 1 and 4).

(i) Probe concentration
Optimal probe concentration is difficult to predict but the criterion should be that which gives greatest signal-to-noise ratio: since background is linearly related to probe concentration, it is best to use the lowest concentration required to saturate target nucleic acids. In our experience, the optimal concentration of cRNA ranges from 5×10^3 to 5×10^6 cpm per section.

(ii) Probe length
The best length for cRNA probes has been shown to be between 50 and 200 nucleotides. Mild hydrolysis can be applied to longer probes prior to hybridization. These penetrate tissue adequately, using standard fixation and permeabilization methods.

(iii) Stringency
Stringency and washing conditions are important when performing *in situ* hybridization (Tecott *et al.* 1987). Stringency refers to the degree to which reaction conditions favour the dissociation of nucleic acid duplexes and may be enhanced, for instance, by increasing temperature, decreasing salt concentration, and increasing formamide concentration; duplexes with high homology withstand high stringency conditions better than duplexes with low homology. Since hybridizations are generally performed at relatively low stringency ($T_m - 25\,°C$), non-specific interactions with less homologous strands may occur, hence post-hybridization washings at increasing stringency are necessary to induce dissociation of imperfect hybrids.

(iv) Stability of hybrids
Factors that enhance hybrid stability include high ionic strength and salt concentration and a high percentage of GC base pairs. By contrast, formamide disrupts hydrogen bonds and hence has a destabilizing effect. However, formamide is included in most *in situ* hybridization reactions as a

means of preventing the association of non-homologous strands at the relatively low temperatures required to maintain adequate tissue morphology.

In situ hybridization is performed under low stringency conditions, at an optimal temperature of T_m −25 °C. However, because of the increased stability of RNA hybrids, *in situ* hybridization using cRNA probes is carried out at higher hybridization temperatures (by about 10–15 °C) than for other probes of comparable length and GC content (Cox *et al.* 1984). Hybridization is generally performed in the dark because ionization of formamide is enhanced by light.

(v) Kinetics

cRNA probes have been shown to produce a stronger signal than double-stranded cDNA probes. Cox *et al.* (1984) compared the signals resulting from hybridization with cDNA and cRNA and found an eightfold greater signal with single-stranded RNA probes. They have suggested that the re-annealing of the cDNA in solution quickly removes most of the probe from the hybridization reaction.

7. Post-hybridization treatments

Post-hybridization washes at increasing stringencies ensures dissociation of imperfect hybrids, since the hybridization step is performed under low-stringency conditions that permit non-specific adherence of probe molecules to various tissue components and background enhancement. RNA probes in particular tend to have high levels of non-specific binding, which are most effectively reduced by post-hybridization treatment with RNAses. RNAses, used under appropriate conditions, selectively remove non-base-paired RNA from tissue sections, so that the background may be greatly reduced with little loss of signal. The acceptable level of background may differ depending on the length of the intended exposure; if a long exposure is planned, more extensive washing may be required. A series of washes of increasing stringency is typically performed by varying the salt concentration, temperature, or formamide concentration (see protocol). The stringency required for an optimal signal-to-noise ratio must be determined empirically (see Chapter 12).

8. Detection system

(i) Isotopic probes (X-ray film-emulsion dipping)

Autoradiography is the most commonly used method for visualization of isotopically labelled bound probe in tissue sections or in cell cultures. Depending on the degree of sensitivity required, the slides can be apposed directly to X-ray film or dipped into molten photographic emulsion. The first method, which requires 2–3 days exposure, affords a quick means of visualizing the isotopically labelled probe and is therefore useful for screening. Once the success of the hybridization has been evaluated, the slides can subsequently be dipped into emulsion. However, the film method is of use only when the nuclei acids of interest are present in high copy numbers in an anatomically defined area, e.g. prolactin in the pituitary gland. The power of *in situ* hybridization techinques, however, is derived from the ability to localize specific genes at the cellular level and in order to do this a dipping method is necessary. Although this method is sensitive, its disadvantage is that extended exposure times are generally required, but again these depend on the abundance of nuclei acids present. Exposure times are usually within the following range: ^{3}H, 6–18 weeks, ^{35}S and ^{32}P, 3–15 days. Care should be taken to find the optimal exposure time for each experiment since, although longer exposure times will provide a greater signal, background levels will increase also. Furthermore, the degree of specific labelling is not proportional to the length of exposure time and an end-point will be reached at a certain time when no increase in the signal is observed.

(ii) Non-isotopic probes

Detection of hybridization reaction using non-isotopic probes can be done by a variety of methods (see Chapter 4) including the use of avidin, antibodies to biotin, or labelling the probes with photobiotin (Forster *et al.* 1985). For enzymic detection systems using 3′,3-diaminobenzidine (DAB) as the chromogen, the final reaction product can be enhanced using silver ions (Amersham DAB enhancement kit) or alternatively by methods commonly employed in immunocytochemistry, such as the nickel–DAB–nascent hydrogen peroxide enhancement method (Hsu and Soban 1982). The use of immunogold staining methods for detection of biotin residues, followed by enhancement with silver ions also proves to be a sensitive means of visualization (Springall *et al.* 1984).

9. Determination of specificity

It is essential to verify the specificity of hybridization signals by performing appropriate controls (Tecott *et al.* 1987). Specific hybridization may be easily confused with unwanted hybrid formations between the probe and weakly homologous sequences or with non-specific interactions between probe and non-nucleic acid tissue components. A number of criteria may be used to assess the specificity of hybridization: pretreatment of the tissue with RNAses, use of heterologous (unrelated) probes, use of multiple probes for a single target, melting point analysis of hybrids for *in situ* hybridization, or correlation with immunocytochemistry. A combination of these strategies will provide a strong indication of specificity.

10. Quantification

Many factors must be carefully considered and controlled if quantitative data are to be collected: section thickness, nucleic acid retention, consistency of hybridization, exposure and development conditions, and thickness and uniformity of emulsion-coating. Essential for acquisition of quantitative data is the use of standard curves, which permit the correlation of the grain density resulting from a particular procedure with the absolute levels of nucleic acid (for details, see Chapter 7).

Specific illustrative examples

In this section, three examples of *in situ* hybridization using cRNA probes will be given. These examples have been chosen in order to illustrate three separate uses of *in situ* hybridization, namely: (a) peptide gene expression in unexpected locations; (b) *in situ* hybridization for assessment of endocrine cell activity; and (c) the usefulness of mRNA technology when the translated protein is stored in low quantity.

(a) Peptide gene expression in unexpected locations
(Gibson *et al.* 1988)

Calcitonin gene-related peptide (CGRP) is a 37-amino-acid peptide encoded in the same gene as calcitonin. By contrast with calcitonin, the mRNA for CGRP is abundantly expressed in nervous tissue, in particular the sensory nervous system, and poorly expressed in the thyroid. Using specific antibodies to CGRP it was possible to detect immunostained neurones not only in the sensory nervous system but also, unexpectedly, in motoneurones. In order to rule out some unspecific reaction in motoneurones (e.g. false impression due to immunostained nerve terminals impinging onto motoneurones, or unspecific peptide uptake), *in situ* hybridization was carried out using sections of human and rat spinal cord and dorsal root ganglia fixed in 4% paraformaldehyde hybridized with sense and antisense RNA probes radioactively labelled (Fig. 6.2(A)). Primary sensory neurones were heavily labelled, as were motoneurones of the ventral horn of the spinal cord, in particular of man (Fig. 6.2(B, C)).

These findings are exciting in view of the immunostaining of CGRP in motor end plates and of the reported long-term trophic actions of CGRP at the neuromuscular junction.

(b) *In situ* hybridization for assessment of endocrine cell activity

The model for this experiment was the rat pituitary gland, subjected to a variety of hormonal changes (Steel *et al.* 1988). Prolactin mRNA levels in

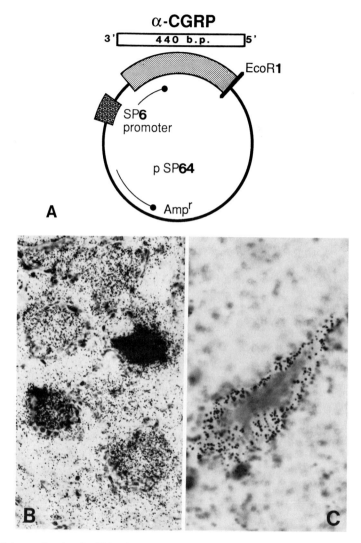

Fig. 6.2. (A) Scheme showing the SP64 plasmid used to generate rat α-CGRP cRNA probes. (B, C) Autoradiograms following hybridization with ^{35}S antisense CGRP probes: (B) primary sensory neurones in the rat lumbar dorsal root ganglion; (C) a motoneurone in the lumbar ventral spinal cord of man. *Magnification B × 300, C × 800.*

the pituitaries of normal, pregnant, lactating, and gonadectomized animals were analysed using cRNA probes (Fig. 6.3(A–D)).

Marked differences in labelling intensity and frequency of labelled cells were noted. mRNA for prolactin was greatly reduced during pregnancy when compared with control animals that show strong labelling in numerous distinct cells. Gonadectomized animals, by contrast, showed

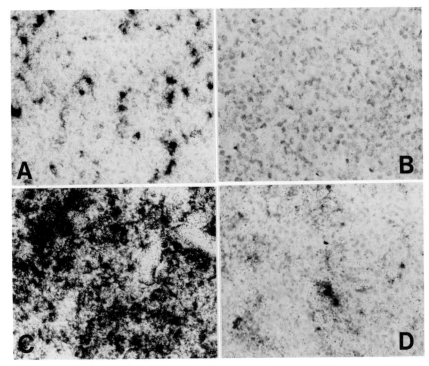

Fig. 6.3. Sections of female rat anterior pituitaries hybridized with ^{32}P-labelled cRNA probe encoding prolactin mRNA: (A) normal gland; (B) pregnant; (C) lactating; (D) ovariectomized.

strong labelling only in scattered cells, suggesting that ovariectomy turns off the message only in a subpopulation of prolactin-producing cells, whereas in pregnancy a generalized switch-off of gene expression occurs. Anterior pituitaries of lactating rats showed a marked increase in message throughout the whole gland.

(c) Strong signal for bombesin mRNA in neuroendocrine tumours containing low quantity of peptide

The example in this case is the small-cell carcinoma of the lung (Hamid *et al.* 1989). This neuroendocrine tumour is characterized by its rapid growth and the frequent production of a growth-promoting factor termed bombesin or gastrin-releasing factor. Ultrastructurally these tumours contain very scattered secretory granules, suggestive of poor peptide storage.

The question is thus posed: Is the mRNA highly expressed, and hence is *in situ* hybridization for bombesin a useful method for the functional characterization of these tumours at the cellular level? We have used cRNA techniques and bombesin probes to systematically hybridize sections of small-cell carcinoma and compare the results with immunocytochemistry

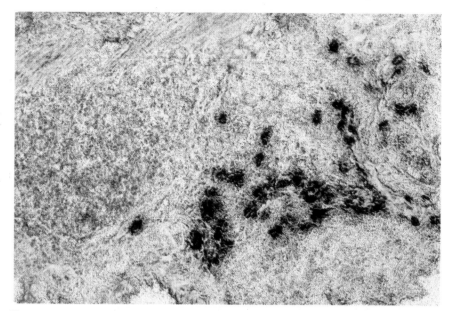

Fig. 6.4. *In situ* hybridization of a section of a small-cell carcinoma of the lung showing positive signal with ^{32}P-labelled cRNA probe encoding preproGRP (mammalian bombesin).

for bombesin. *In situ* hybridization showed a high expression of the bombesin gene, the technique frequently being more helpful than immunocytochemistry and electron microscopy (Fig. 6.4).

Conclusion

cRNA techniques for the investigation of the mRNAs coding for regulatory peptides of the 'diffuse neuroendocrine system' have provided very useful insights into the basic mechanisms of peptide production. In many situations they have become a very good adjunct to or even better than the existing useful methods of immunocytochemistry and electron microscopy.

References

Cox, K. H., DeLeon, D. V., Angerer, L. M., and Angerer, R. C. (1984). Detection of mRNAs in sea urchin embryos by *in situ* hybridization using asymmetric RNA probes. *Develop. Biol.*, **101**, 485–502.

Denny, P., Hamid, Q., Krause, J. E., Polak, J. M., and Legon, S. (1988). Oligoriboprobes: Tools for *in situ* hybridisation. *Histochemistry*, **89**, 481–3.

Forster, A. C., McInnes, J. L., Skingle, D. C., and Symons, R. H. (1985). Nonradioactive hybridization probes prepared by the chemical labeling of DNA and RNA with a novel reagent, photobiotin. *Nucl. Acids. Res.*, **13**, 745–61.

Gibson, S. J., Polak, J. M., Giaid, A., Hamid, Q. A., Kar, S., Jones, P. M., Denny, P., Legon, S., Amara, S. G., Craig, R. K., Bloom, S. R., Penketh, R. J. A., Rodek, C.,

Ibrahim, N. B. N., and Dawson, A. (1988). Calcitonin gene-related peptide messenger RNA is expressed in sensory neurones of the dorsal root ganglia and also in spinal motoneurones in man and rat. *Neurosci. Lett.*, **91**, 283–8.

Hamid, Q. A., Bishop, A. E., Springall, D. R., Adams, C., Giaid, G., Denny, P., Ghatei, M., Legon, S., Cuttitta, F., Rode, J., Spindel, E., Bloom, S. R., and Polak, J. M. (1989). Detection of human pro-bombesin mRNA in neuroendocrine (small cell) carcinoma of the lung: *in situ* hybridization with cRNA probes. *Cancer*, **63**, 266–71.

Hsu, S.-M. and Soban, E. (1982). Color modification of the diaminobenzidine (DAB) precipitation by metallic ions and its application to double immunohistochemistry. *J. Histochem. Cytochem.*, **30**, 1079–82.

Lawrence, J. B. and Singer, R. H. (1985). Quantitative analysis of *in situ* hybridization methods for the detection of actin gene expression. *Nucl. Acids Res.*, **15**, 1777–99.

Springall, D. R., Hacker, G. W., Grimelius, G. W., and Polak, J. M. (1984). The potential of the immunogold-silver staining method for paraffin sections. *Histochemistry*, **81**, 603–8.

Steel, J. H., Hamid, Q., Van Noorden, S., Jones, P., Burrin, J., Legon, S., Bloom, S. R., and Polak, J. M. (1988). Combined use of *in situ* hybridisation and immunocytochemistry for the investigation of prolactin gene expression in immature pubertal, pregnant, lactating and ovariectomised rats. *Histochemistry*, **89**, 75–80.

Tecott, L. H., Eberwine, J. H., Barchas, J. D., and Valentino, K. L. (1987). Methodological consideration in the utilization of *in situ* hybridisation. In *In situ hybridization: applications to neurobiology* (ed. K. L. Valentino, J. H. Eberwine, and J. D. Barchas), Oxford University Press, New York.

Wetmur, J. G., Ruyechan, W. T., and Douthart, R. J. (1981). Denaturation and renaturation of Penicillium chrysogenum mycophage double-stranded ribonucleic acid in tetraalkylammonium salt solutions. *Biochemistry*, **20**, 2999–3002.

Wolf, S., Quaas, R., Hahn, U., and Wittig, B. (1987). Synthesis of highly radioactively labelled RNA hybridisation probes from synthetic single-stranded DNA oligonucleotides. *Nucl. Acids Res.*, **15**, 858.

Appendix 6.1

In situ hybridization protocols for detection of mRNA in tissue sections/cultures

A. Transcription protocol for cRNA probes

1. Add together in the following order:

- 2.0 μl 5 × transcription buffer [(0.2 M Tris HCl (pH 7.5), 30 mM $MgCl_2$, 10 mM spermidine]
- 1.0 μl 100 mM dithiothreitol

- 0.4 μl RNasin (human placental ribonuclease inhibitor) (1 unit/μl)
- 2.0 μl unlabelled nucleotide mixture: 2.5 mM each of adenosine, guanine, and uridine triphosphates
- 1.2 μl 100 μM cytidine triphosphate
- 0.5–0.75 μl linearized plasmid template DNA (1 mg/ml) in water or Tris EDTA buffer
- 2.5 μl cytidine triphosphate radiolabelled with either α ^{35}S or ^{32}P (10 mCi/ml)
- 0.5–1.0 μl SP6, T7 or T3 riboprobe polymerase (10 units/μl)

2. Incubate for 1.5 h at 37–40 °C.

3. To terminate transcription, add 1 μl DNase (1 μg/μl) and 1 μl RNasin for 10–20 min at 37 °C.

4. Then add 1 μl total RNA (10 μg/μl), 175 μl DEPC-treated H_2O, 5 μl NaCl (4 M).

5. Extract with an equal volume (200 μl) of phenol/chloroform (1:1 v/v) saturated with H_2O or Tris/EDTA buffer. Vortex thoroughly and microfuge for 5 min.

6. Separate aqueous layer (200 μl) and extract with an equal volume of chloroform. Vortex and microfuge for 2–3 min.

7. To the aqueous layer add 100 μl 7 M ammonium acetate (2.5 M final concentration) and 750 μl absolute ethanol (-20 °C). Vortex and leave to precipitate for 10–16 h at -20 °C.

8. Microfuge for 30 min. Discard the supernatant. Vacuum dry the pellet and dissolve in 20 μl DEPC-treated water. Remove 1 μl for assessment of incorporation of radioactivity.

B. Prehybridization treatment

1. Adhere sections/cultures to pretreated slides and bake overnight at 43 °C.

2. Rehydrate in phosphate-buffered saline (PBS) (0.1 M, pH 7.2) for 5–10 min.

3. Immerse in 0.1 M glycine in PBS for 5 min.

4. Soak slides in 0.3% Triton X-100 in PBS for 10–15 min to permeabilize.

5. Wash thoroughly with PBS 3 × 5 min.

6. Incubate with proteinase K (1 μg/ml) in Tris HCl (0.1 M, pH 8.0) and EDTA (50 mM, pH 8) for 20 min at 37 °C.

7. Stop reaction by immersion in freshly made 4% paraformaldehyde in PBS (0.1 M, pH 7.2).

8. Acetylate with acetic anhydride (0.25% v/v) in triethanolamine (0.1 M, pH 8.0) for 10 min.

9. Prehybridize in formamide (50% v/v) in $4 \times$ SSC at 37 °C for 15–45 min.

C. Hybridization

    1. Drain excess formamide solution from slides and apply 20 μl of hybridization mixture at 42 °C containing radiolabelled probe ($5 \times 10^3 - 10^6$ cpm per section or culture well).

    2. Cover the sections/cultures with dimethylsilane-coated coverslips and incubate in a humid atmosphere at 42 °C for 12–18 h.

D. Post-hybridization washing

    1. Immerse slides in $4 \times$ SSC to remove coverslips.

    2. Wash slides with gentle agitation in $4 \times$ SSC at 42 °C, 3×20 min.

    3. Immerse in RNAse A (20 μl/ml) in NaCl (0.5 M), Tris-HCl (pH 8.0) and EDTA (1 mM, pH 8.0) at 42 °C for 30 min.

    4. Wash sections in descending concentrations of SSC; $2 \times$ SSC, $0.1 \times$ SSC and 0.05 SSC, 30 min each at 42 °C.

    5. Dehydrate with 70%, 90% and in two changes of 100% ethanol containing 0.3 M ammonium acetate 10 min each at room temperature. Air dry.

7

Quantification of radioactive mRNA
in situ hybridization signals

ANTHONY P. DAVENPORT AND DEREK J. NUNEZ

Introduction

In situ hybridization is widely used for detecting DNA and RNA sequences in normal and pathological tissues by the use of complementary DNA and RNA probes that have been labelled with a radioactive molecule. After the process of hybridization is complete, sections are washed with increasing stringency, dried and apposed to radiation-sensitive emulsion to record by autoradiography the spatial distribution of the probes within the thin (10–25 μm) sections of tissue. Quantification of non-isotopic signals has been employed for chromosomal gene assignment (see Chapter 10) but not in tissue localization of nucleic acids. This chapter, therefore, only considers quantification of radioactive hybridization signals *in situ* hybridization in tissue sections.

Two techniques are employed, depending on the level of resolution required. Using micro-autoradiography, radioactivity is detected when viewed under the compound microscope as the pattern of individual silver grains in a thin layer of nuclear emulsion lying above the tissue. This is applied either by dipping the slides into liquid emulsion or by apposition to emulsion-coated coverslips. Using tritium-labelled oligonucleotide probes, it is possible to visualize radioactivity associated with a single cell. Radioactivity and therefore the amount of probe hybridized has been assessed by counting the number of silver grains per cell or unit area lying directly above the specimen (Shivers *et al.* 1986; Uhl *et al.* 1986; Young and Kuhar 1986). Measurement of grain density can be difficult to make with precision because numerous factors affect the response of the emulsion to radiation, including the characteristics of the radiation source (for detailed discussion, see Rogers 1979), the length of exposure, and conditions of development. For example, isotopes such as ^{35}S and ^{32}P, emit β particles of sufficiently high energy to pass through the emulsion and, under these conditions, grain densities may vary with emulsion thickness (Young and Kuhar 1986). This problem can be avoided by the use of tritium to label the probes, since this isotope produces particles with a maximum range in emulsion of about

95

2 μm. Provided the emulsion is 10–15 μm thick, all β particles emitted in the direction of the emulsion should be detected. However, variation in tissue density can result in differential self-absorption of the low-energy particles emitted by tritium, producing errors in measurement of the concentration of radioactivity and therefore of the amount of probe hybridized in a single cell.

When single-cell resolution is not required, an alternative approach is to use macro-autoradiography, where many of these factors can be controlled for or avoided by co-exposing the radiolabelled sections to a thin layer of emulsion coated onto a polyester film base with a set of calibrated standards that closely resembles the experimental source. At the end of the exposure period, the film is separated from the slides, and after development the resulting pattern of optical densities within the autoradiograms can be quantified by comparison with a standard curve generated from the calibrated radioactive scale using computer-assisted densitometry.

The main advantage of this method for quantifying mRNA levels compared to solution hybridization (which involves extraction of nucleic acids from tissue homogenates) is that the morphology of the tissue is retained, permitting the amount of radioactivity to be determined in areas as small as 500–1000 μm^2, which is equivalent to 5–10 ng of tissue in a 10-μm thick section. In complex structures such as the brain, many discrete regions can be resolved in a single coronal section without preselection of the tissue to be analysed. Secondly, autoradiography can detect lower levels of radioactivity than methods based on spectrophotometry. In addition, *in situ* hybridization can be combined with other techniques such as receptor autoradiography and immunocytochemistry. Thus, for a neuropeptide, changes in levels of the neurotransmitter, its receptors, and messenger RNA coding for the neurotransmitter, can be measured in consecutive 10-μm thick sections from the same tissue.

This chapter will concentrate on describing one approach towards the quantification of mRNA by computer-assisted image analysis and will recommend some strategies for validating the technique and avoiding potential errors. For a more detailed discussion of *in situ* hybridization, see, for example, reviews by McCabe *et al.* (1986), Penschow *et al.* (1987), Valentino *et al.* (1987), and Lewis *et al.* (1988). For quantitative autoradiography and image analysis, see Baskin and Dorsa (1986), Eilbert (1986), Hall *et al.* (1986), McEachron *et al.* (1986), and Davenport *et al.* (1988a,b) and Chapter 3.

Autoradiography

Following *in situ* hybridization, the dry slides are securely mounted to rigid board using adhesive tape to prevent movement and placed in a light-tight X-ray cassette. Up to 64 slides including standards can be apposed at the same time to 35×43 cm films. In complete darkness (or with minimal exposure to red safelight (Kodak Wratten 6B or equivalent)), sections are apposed to a radiation-sensitive film. Single emulsion films such as Hyperfilm βmax (Amersham International) are recommended (Davenport *et al.* 1988*a,b*) for ^{35}S, ^{32}P, and ^{125}I; Hyperfilm ^{3}H, which lacks the anti-abrasive layer, is useful for the detection of ^{3}H and ^{125}I isotopes.

At the end of the exposure, films are manually developed in Kodak D-19 developer for 3 min at 17 °C with intermittent agitation, rinsed for 30 s in deionized water and fixed in Kodak Kodafix for 5 min at 20 °C. Films are washed for at least 15 min in running filtered water, rinsed in deionized water, and suspended in a drying cabinet. All procedures should be carried out in total darkness or with minimal exposure to a red safelight to monitor the progress of development.

Minimizing errors in detecting radioactivity

1. Slides bearing replicate sections should be positioned throughout the area occupied by the film, to minimize possible variation in the thickness of the emulsion.
2. Sections from control and experimental animals should be mounted on the same slide.
3. Ensure that 'X-ray' cassettes produce uniform pressure over the film.
4. The use of intensifying screens to increase the autoradiographic efficiency by converting ionizing radiations to light should be avoided.

Quantification of autoradiograms: Computer-assisted image analysis

The resulting autoradiograms are analysed by measuring diffuse integrated optical density using a computer-assisted image analysis system (Fig. 7.1). We use a Quantimet 970 (Cambridge Instruments) employing a computer program developed for quantitative autoradiography (Davenport *et al.* 1988*a,b*). A wide range of image analysers are available but in order to carry out densitometry the machine should ideally be equipped with a shading corrector and be able to digitize images into an array of at least 500×500 picture points with 64 or 256 grey levels. The ability to digitally subtract stored images increases the speed of analysis.

Using the Quantimet system, autoradiograms are illuminated by reflected white light and the image is captured by a Chalnicon videoscanner equipped with a zoom lens and digitized into an array of 630 000 image points each with a grey value in the range 0–255. Once captured, the operator can manipulate images presented on a high-resolution monitor using a digitablet employing a hand-operated cursor and an alphanumeric keyboard. Images can be colour-coded by reference to a previously

Image and Data Recording	Image Display	Image Processing	Image Acquisition

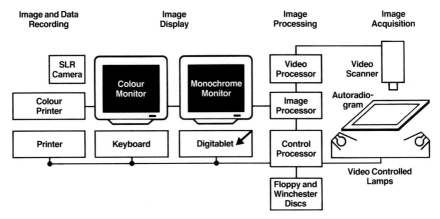

Fig. 7.1. Schematic diagram of a modern image analyser. (From Davenport *et al.* 1988b.)

measured standard curve and are recorded directly from the screen onto 35-mm film. Programs, data, and images can be stored on Winchester drives or floppy discs.

The zoom lens is altered to produce a measuring field appropriate to the size of the autoradiogram to be analysed and the shading corrector is set by imaging a blank area of the film. The shading corrector compensates for variation in illumination and the light-transmitting properties of the scanner. White level (100 per cent transmission) is set by imaging a blank area of the film and the zero offset is eliminated by interrupting the light path with an opaque object. Finally, the system is calibrated against neutral-density filters to convert grey levels into optical densities and number of pixels per unit area calculated by means of a measuring box. The image analyser should be calibrated for densitometry each time the conditions are changed, such as alteration of the magnification or use of a different film.

Constructing a standard curve from calibrated standards

To measure the optical densities of a set of standards, the threshold is set to detect each standard in turn and the cursor is used to draw around the detected image. The integrated optical density for each standard together with the area is determined. Radioactivity measured by liquid scintillation counting is then divided by the area to calculate radioactivity in dpm/mm^2 which is stored on disc.

The relationship between film optical density and tissue radioactivity may vary as a result of factors including the characteristics of the radiation source, the type of emulsion, conditions of exposure, and method of development. A number of mathematical transformations have been used to describe this relationship, but we prefer to use a natural log plot of optical density versus radioactivity to give a linear relationship as shown in Fig. 7.2 for a series of ^{35}S tissue paste standards apposed to Hyperfilm βmax.

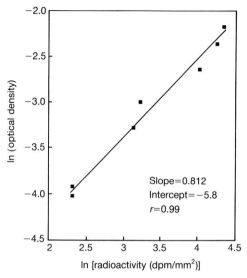

Slope=0.812
Intercept=−5.8
r=0.99

Fig. 7.2. An example of a natural-log plot of optical density versus radioactivity (dpm/mm^2) for 10-μm thick ^{35}S tissue paste standards exposed directly to radiation-sensitive film for 3 days. Each value represents the mean of three determinations. Standard errors were less than 3 per cent of the mean.

Minimizing errors in quantifying autoradiograms

1. Standards of the same isotope used to label the tissue sections should be co-exposed with each film.

2. Hyperfilm [^{3}H] and βmax rapidly approach saturation at an optical density of one (Davenport and Hall 1988). At this point, increases in radioactivity result in little or no increase in optical density. It is important to ensure that when using probes labelled with ^{35}S, ^{32}P or ^{125}I, areas of the film are not so heavily irradiated so as to produce images approaching the saturation of the film. Films should be re-apposed where necessary.

3. The upper and lower limits for making accurate measurements for each type of film and image-analyser should be determined. Using the Quantimet, variation in optical densities of less than 5 per cent are obtained within the range 0.1–0.8. Ideally, tissue sections should be exposed to produce optical density values falling within this range.

4. Check for regional variation in absorption of beta particles emitted by tritium, which could result in varying autoradiographic efficiency over a

tissue section (for discussion, see Davenport *et al.* 1988*a,b*). No differences in adsorption have been observed using isotopes emitting β particles of higher energy than tritium.

Measuring the density of probe hybridization to tissue sections

An example of a film-based autoradiogram is shown in Fig. 7.3(A), in which a 42-mer ^{35}S-oligodeoxyribonucleotide probe (complementary to ANF mRNA sequence coding for pro-ANF 103-116) has been used to localize messenger RNA coding for atrial natriuretic factor in a longitudinal section of rat heart (Nunez *et al.* 1989). A cursor is used to draw within a defined anatomical area such as the atria. The computer isolates this area and measures the integrated optical density. When all measurements have been completed for this particular autoradiogram, the threshold for detecting the section is then increased to produce a template with which to align the autoradiographic image of an adjacent section treated with RNAse (Fig. 7.3(B)) prior to the hybridization assay. The image of the RNase-treated section is digitally subtracted from the first to obtain a new image of the 'specific' hybridization of the probe. The amount of radioactivity and, therefore, the amount of probe specifically hybridized to the structure is calculated in attomoles per square millimetre by interpolation from the previously generated standards curve (Fig. 7.2). Molar quantities of probe hybridized per unit area are obtained using the following equation:

$$\text{Probe bound (amol/mm}^2) = \text{dpm/mm}^2 \times (\text{SA})^{-1}$$

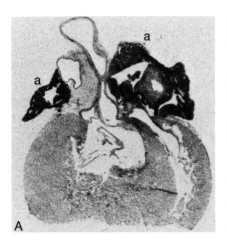 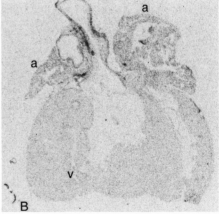

Fig. 7.3. (A) Autoradiographic localization of a ^{35}S-labelled oligonucleotide probe to atrial natriuretic factor (ANF) mRNA in a longitudinal section from an adult rat heart. Following *in situ* hybridization, the section was washed, dried, and apposed to radiation-sensitive film. (B) An adjacent section to that used to produce (A) was incubated for 30 min with RNase (100 μg/ml) prior to *in situ* hybridization in order to assess non-specific binding of the ANF mRNA probe. Atria (a); ventricle (v).

where SA is the specific activity of the probe at the mid-point between exposure and development of the film. In order to obtain an accurate measurement of the specific activity, labelling protocols must be modified to ensure the addition of a single radioactive reporter molecule. Since standards are used repeatedly, the radioactive content should be corrected for decay from the date when they were counted and the mid-point between apposing and developing the film.

Application of quantitative *in situ* hybridization

The application of the technique will be illustrated with a study in which we have measured changes in ANF mRNA in the adult rat heart produced by mineralocorticoid/saline treatment and also compared adult and neonatal cardiac tissues to investigate the modulation of ANF secretion through alteration of gene expression.

Figure 7.4(a) shows the amount of probe hybridized to adult hearts from control and saline/deoxycortone-treated rats. In both cases, there is no significant difference between ANF mRNA probe hybridization to left and right atria. However, the amount of probe hybridized to the atria in the controls was about 70 times higher compared to the ventricles. In the mineralocorticoid-treated animals, the amount of probe hybridized to atria and ventricles was greater than in the controls but the ventricles showed a larger increase.

In Fig. 7.5 we have compared specific ANF probe hybridization to ANF mRNA in neonate hearts. Again, there was no significant difference between left and right atria, and levels were 30 times lower in the ventricles. Comparing these results with the adult hearts, in neonatal atria the amount hybridized is three times that in the adult chamber. These results suggest that, in neonates, enhanced expression of the ANF gene is occurring throughout the heart, especially in the ventricles.

Validation of the technique

In order to carry out accurate and reproducible quantification of mRNA in tissue sections, it is important to validate the method by considering the following questions: (1) what is the specificity of the assay? (2) what is its sensitivity? (3) what is its reproducibility? and (4) what is its accuracy?

Specificity

At present, there is no generally accepted single criterion for assessing the specificity of a particular probe. Therefore, a combination of techniques must be used. As a first step, the probe should be checked for the lack of significant complementarity to other nucleic acids submitted to the EMBL and Genbank nucleic acid data bases.

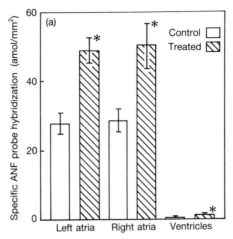

Fig. 7.4. (a) Specific hybridization of the 3′ end-labelled ANF mRNA probe to adult control (open bars) and saline/deoxycortone-treated rats (hatched bars). In the control animals there was no significant difference between left and right atria. The levels of ANF mRNA probe binding in atria and ventricles were significantly elevated in the saline/mineralocorticoid-treated rats. Each value represents the mean of 12 determinations ± standard error of the mean. (From Nunez *et al.* 1989, with permission.)

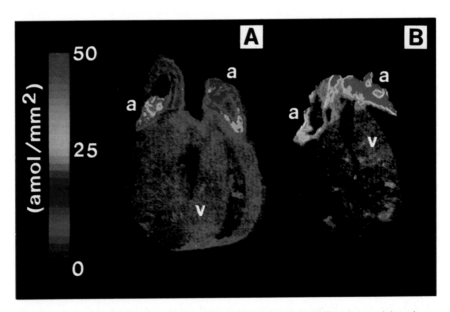

Fig. 7.4. (b) Colour-coded images of the specific hybridization of the ANF probe to adult rat heart. In the control rat heart (A), the atria (a) bind the highest amount of probe (30 amol/mm^2, shown in red) with detectable but lower binding in the ventricles (v). In the mineralocorticoid-treated, saline-loaded rat (B), levels in the atria and ventricles are elevated compared to controls.

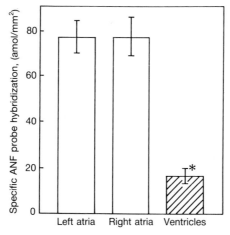

Fig. 7.5. Specific hybridization of the 3′ end-labelled ANF mRNA probe to neonatal rat hearts. There was no significant difference between probe hybridization to left and right atria. Significantly lower levels were detected in the ventricles. Each value represents the mean of 16 determinations ± standard error of the mean. (From Nunez *et al*. 1989.)

Northern analysis of cellular RNA

Total cellular RNA from the tissue under investigation should be subjected to standard northern analysis using the labelled oligonucleotide probe. After hybridization and washing under conditions similar to those used in the *in situ* hybridization protocol, there should be a single band of radio-activity corresponding to the expected position of the mRNA under investigation. No bands should be seen at the positions expected for ribosomal or transfer RNA, indicating that the probe does not interact significantly with these nucleic acids within the tissue section. It should be borne in mind that tRNA (and 5s rRNA) may remain solubilized during acetate/ethanol precipitation so that the amounts electrophoresed and blotted may be too low to exclude significant probe binding. However, the predominantly double-stranded nature of these polynucleotides suggests that non-specific hybridization to them should be low.

Measurement of non-specific hybridization

In order to obtain a measure of the hybridization of the probe to proteins and other non-single-stranded mRNA species, adjacent sections of tissue should be pre-treated for 30 min at 37 °C with a solution (100 μg/ml) of RNase (Sigma) in 0.5 NTE enzyme buffer (10 mM Tris, 0.5 M NaCl, 1 mM EDTA, pH 8) previously boiled to destroy DNase activity. The sections are then incubated for a further 30 min in the buffer alone. At the concentration of RNase A employed, it is expected that all single-stranded RNA species are cleaved and are therefore unavailable for hybridization.

Double-stranded RNA is relatively resistant to this enzyme and would be expected to remain in the tissue sections, but this should not represent a significant proportion of each messenger RNA molecule. We have found that non-specific hybridization of probes is usually low and can be digitally subtracted from the autoradiographic image of the total section to give the amount of specific probe hybridization as previously described (Brown *et al.* 1988*a,b*; Nunez *et al.* 1989).

Comparison with other probes

Oligodeoxynucleotides can be synthesized that are complementary to nucleotide sequences upstream of the postulated initiation site for the transcription of the mRNA of interest. Sense probes can also be made. The latter have an identical $G + C$ content and should be of the same length. Theoretically, none of the latter, when labelled, should hybridize to the tissue; any binding to the sections is non-specific and gives a measure of the general 'stickiness' of polynucleotides. This may reveal any unusual DNA binding properties of a particular tissue.

We are now routinely using a battery of probes to a particular mRNA species to document precise co-localization. Furthermore, quantification of hybridization should reveal parallel and identical changes in different pathophysiological states.

The use of thermal stability may not be a good criterion for the validation of probe specificity, as the melting temperatures of probes to tissue sections and blots are relatively insensitive to change in the size of the probe (from 40-mer to 1 kb) suggesting that mismatches may not be easy to detect by this method (B. J. Morris, personal communication, Chapters 5 and 12).

Determining optimum probe concentrations

In order to ensure that there is an excess of probe during hybridization, the effect of varying the probe concentration should be investigated. The result of such an experiment is shown in Fig. 7.6 using the same ANF probe labelled either at the 5′ end (shown in circles) or the 3′ end (shown as squares).

For both probes, there was a linear relationship between probe concentration (in the range 0.5–8 fmol/μl) when applied in a volume of 50 μl and the amount of oligonucleotide hybridized to adult rat atria. There was no significant difference between the slopes of these two lines. The amount of probe hybridized reached a plateau above 8 fmol/μl, suggesting that this quantity of probe represents the amount needed to saturate the hybridization sites in the tissue sections and also indicates that there are a finite number of these sites, unlike the situation expected for unsaturable, non-specific binding. A concentration of probe should be selected for subsequent experiments in excess of this level (Nunez *et al.* 1989).

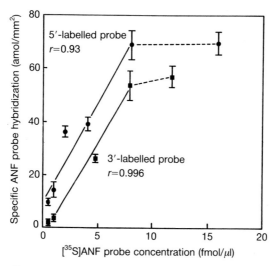

Fig. 7.6. The effect of increasing the concentration of two ^{35}S-labelled ANF mRNA probes applied to 10-μm cryostat sections of adult rat atria. Probe 1 was labelled with 5'-γ-[^{35}S]dATP at the 5' position with T4 polynucleotidyl kinase. Probe 2 was labelled at the 3' position with 5'-α-[^{35}S]ddATP using terminal transferase. When the data for the highest concentration were omitted, for both probes there was a linear correlation between specific probe hybridization and probe concentration. There was no significant difference between the slopes of the two lines. The amount of probe hybridization reached a plateau after an inflection point at a concentration of 8 fmol/μl when applied in a volume of 50 μl. Each value represents the mean of four determinations $\pm$ the standard error of the mean. (From Nunez *et al.* 1989.)

Section thickness

The effect of increasing cryostat section thickness on the amount of 5' ^{35}S-labelled ANF mRNA probe bound to control adult rat atrium and ventricle is shown in Fig. 7.7. There was a linear relationship between section thickness in the range 5–30 μm and the amount of specific probe hybridization for both atria and ventricles, as would be expected because increasing the section thickness increases the total RNA per unit area. This suggests there are no barriers to the penetration of the probe.

Sensitivity

A measure of the sensitivity of an *in situ* hybridization assay can be calculated from the type of data presented in Fig. 7.4, which shows the amount of specific ANF mRNA hybridization to control rat ventricles (0.41 amol/mm^2). The number of cells in a standardized volume of tissue under investigation is calculated. For example, the average myocardial cell volume can be calculated as 125 μm^3, assuming the average cardiac cell dimensions to be about 5 μm $\times$ 5μm $\times$ 5μm. Thus, in control ventricular tissue three copies of the ANF mRNA per cell are detectable by this technique.

In situ *hybridization*

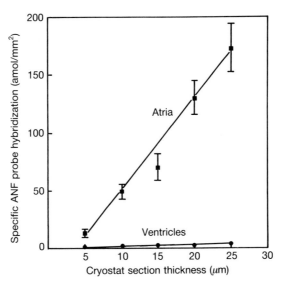

Fig. 7.7. The effect of increasing cryostat section thickness on the amount of the $5'$-^{35}S-labelled ANF mRNA probe bound to adult rat atria and ventricles. There was a linear relationship between the section thickness and the amount of labelled oligonucleotide hybridized. There is a significant difference between the slope of the atrial line and that of the ventricles but no difference between the y intercepts. Each value represents the mean of four determinations ± standard error of the mean. (From Nunez *et al.* 1989.)

An alternative method is to use the information given in Fig. 7.7 for the effect of increasing cryostat section thickness on the amount of ANF mRNA probe bound to control cardiac tissues and in Fig. 7.6 showing the effect of increasing probe concentration. There is no significant difference between the slopes of these lines relating specific probe hybridization to probe concentration and section thickness for adult rat atria. Below the inflection point, doubling the concentration (in a 50 μl volume) or thickness results in a doubling of the amount of probe bound. The results in Figs. 7.6 and 7.7 are therefore equivalent to titration curves of the proportion of specific ANF mRNA in the total RNA pools of atrial and ventricular cells. The slopes of these titration curves can be used to determine transcript prevalence/cell as in solution hybridization.

Thus, ANF transcript number can be assessed from these data by using equation:

$$T = \frac{n}{b} \, ms$$

where T is the number of transcripts/cell, n is Avogadro's number, b the specific activity of the probe, m the slope of the titration curve in units of RNase-digestible binding per mass unit of cellular RNA, and s the mass of

RNA/cardiac cell (approximately 2×10^{-5} μg/cell). This yields results similar to those obtained using the first method, and again indicates that the limit of detection is less than five copies of ANF mRNA/cell. From the ventricular results, our calculations suggest that *in situ* hybridization combined with computer-assisted image analysis of autoradiograms has a sensitivity comparable to that of solution hybridization (< 5 transcripts/cell) and appears to be considerably better than blotting techniques, which have been unable to detect ventricular ANF mRNA. However, we have not assessed the efficiency of hybridization of our probes or quantified the loss of mRNA during section preparation, storage, fixation, and hybridization, but the latter is probably small and contributes to the good reproducibilty that we have found in measuring mRNA using this technique. These problems should be considered in the context of disruption of anatomical integrity through homogenization extraction and transfer losses, and difficulties in determining hybridization efficiency with blotting techniques and a requirement for large cDNAs for solution hybridization.

Precision

The reproducibility of an *in situ* hybridization assay can be assessed by measuring the inter- and intra-assay variability. We (Nunez *et al.* 1989) have found that the intra-assay coefficient of variation (CV) is less than 10 per cent and commonly as low as 3.5 per cent, whereas the interassay CV is less than 10 per cent.

Accuracy

At present, it is difficult to ascertain the accuracy of measuring levels of mRNA in tissue sections following *in situ* hybridization, owing to the lack of a fully validated alternative method for measuring the specific hybridization of oligonucleotide probes. Conventional blotting and solution hybridization techniques, as methods for quantifying specific mRNA, have an accuracy that is not easily quantifiable owing to unknown and perhaps variable factors such as extraction and transfer losses and hybridization efficiency. It is difficult to compare directly results obtained with our technique and those produced by these other methods. However, others have shown that the amounts of RNA detected by solution hybridization are similar to those measured by *in situ* hybridization (ISH) (Angerer and Angerer 1986). This aspect of the validation requires further study.

Conclusions and directions for future research

ISH and computer-assisted image analysis of autoradiograms can provide quantitative information on specific mRNA levels in defined areas of tissue sections. A major advantage of this technique is the ability to subtract non-specific binding of the reporter molecule to a tissue section from the total hybridization to yield quantitative information that could not be obtained by visual assessment of the autoradiograms. We have focused on the measurement of ANF mRNA in the rat heart to exploit a number of advantages, including the relatively large amount of ANF mRNA in the atria (which allows easy detection), and the lower but detectable levels in the ventricles; the latter provides an internal test of probe specificity and a measure of sensitivity. The technique should have wider applicability particularly in situations where preservation of anatomical relationships is important in microscopic areas of tissue; where tissue samples are small (e.g. human biopsies) and transcripts are present at low prevalence.

References

Angerer, L. M. and Angerer, R. C. (1981). Detection of poly A+ RNA in sea urchin eggs and embryos by quantitative *in situ* hybridisation. *Nucl Acids Res.*, **9**, 2819–40.

Baskin, D. G. and Dorsa, D. M. (1986). Quantitative autoradiography and *in vitro* radioligand binding. In *Functional mapping in biology and medicine: computer assisted autoradiography*, (ed. D. L. McEachron), pp. 204–34. Karger, Basel.

Brown, M. J., Davenport, A. P., Emson, P. C., Hall, M. D., and Nunez, D. J. (1988a). Adrenal cells express atrial natriuretic peptide mRNA (ANP mRNA). An 'in-situ' hybridization study. *J. Physiol.*, **398**, 76P.

Brown, M. J., Davenport, A. P., Emson, P. C., Hall, M. D. and Nunez, D. J. (1988b). Measurement of atrial natriuretic peptide messenger RNA (ANP mRNA) in the neonatal rat heart by computer-assisted image analysis of 'in-situ' hybridization (ISH). *J. Physiol.*, **398**, 75P.

Davenport. A. P. and Hall, M. D. (1988). Comparison between brain paste and polymer [125I] standards for quantitative receptor autoradiography. *J. Neurosci. Methods*, **25**, 75–82.

Davenport, A. P., Hill, R. G., and Hughes, J. (1988a). Quantitative analysis of autoradiograms. In *Regulatory peptides*, (ed. J. Polak), pp. 137–53. Birkhauser, Basel.

Davenport, A. P., Beresford, I. J. M., Hall, M. D., Hill, R. G., and Hughes, J. (1988b). Quantitative autoradiography in neuroscience. In *Molecular neuroanatomy*, (ed. Van Leeuwen, F. W., Buijs, R. M., Pool, C. W., and Pach, O.), pp. 121–45. Elsevier, Amsterdam.

Eilbert, J. L. (1986). Quantitative analysis of autoradiographs. In *Functional mapping in biology and medicine: computer assisted autoradiography*, (ed. D. L. McEachron), pp. 122–42. Karger, Basel.

Hall, M. D., Davenport, A. P., and Clark, C. R. (1986). Quantitative receptor autoradiography. *Nature*, **324**, 493–4.

Lewis, M. E., Krause, R. G., and Robert-Lewis, J. M. (1988). Recent developments in the use of synthetic oligonucleotides for *in situ* hybridization histochemistry. *Synapse*, **2**, 308–16.

McCabe, J. T., Morrell, J. I., Richter, D., and Pfaff, D. W. (1986). Localization of neuroendocrinology relevant RNA in brain by *in situ* hybridization. *Frontiers Neuroendocrinol.*, **9**, 149–67.

McEachron, D. L., Adler, N. T. and Tretiak, O. J. (1986). Two views of functional mapping and autoradiography. In *Functional mapping in biology and medicine: computer assisted autoradiography*, (ed. D. L. McEachron), pp. 1–46. Karger, Basel.

Nunez, D. J., Davenport, A. P., Emson, P. C., and Brown, M. J. (1989). A quantitative *in-situ* hybridization method using computer-assisted image analysis. Validation and measurement of atrial natriuretic peptide messenger RNA in the rat heart. *Biochem. J.*, **263**, 121–7.

Penschow, J. D., Haralambidis, P. E., Darling, P. E., Darby, I. A., Wintour, E. M., Tregear, G. W., and Coghlan, J. P. (1987). Hybridization histochemistry. *Experientia*, 741–9.

Rogers, A. W. (1979). *Techniques of autoradiography*, (3rd edn.). Elsevier, Amsterdam.

Shivers, B. D., Harlan, R. E., Romano, G. J., Howells, R. D., and Pfaff, D. W. (1986). Cellular location and regulation of proenkephalin mRNA in rat brain. In *In situ hybridization in brain*, (ed. G. R. Uhl), pp. 3–20. Plenum Press, New York.

Uhl, G. R., Evans, J., Parta, M., Walworth, C., Hill, K., Sasek, C., Voigt, M., and Reppert, S. (1986). Vasopressin and somatostatin mRNA. In *In situ hybridization in brain*, (ed. G. R. Uhl), pp. 21–47. Plenum Press, New York.

Valentino, K. L., Erberwine, J. H., and Barchus, J. D. (eds.) (1987) In *In situ hybridization: applications to neurobiology.* Oxford University Press, New York.

Young, W. S. and Kuhar, M. J. (1986). Quantitative *in situ* hybridization and determination of mRNA content. In *In situ hybridization in brain*, (ed. G. R. Uhl), pp. 245–8. Plenum Press, New York.

Appendix 7.1

Preparation of standards for the quantification of mRNA by *in situ* hybridization

Calibrated radioactive scales are conveniently prepared by mixing increasing quantities of the radionuclide to be quantified with tissue paste derived from the same target tissues as those used in the *in situ* hybridization assay. The method described below for ^{35}S brain paste standards is modified from Davenport and Hall (1988) and can be adapted to prepare standards from non-neural tissue with other radionuclides.

The brain is removed from six adult male rats and 50 μl of silicone oil is added to the tissue to reduce frothing before homogenization. Known amounts of radioactivity in the form of the ^{35}S-labelled nucleotide, such as deoxycytidine 5′-α-[^{35}S]thiotriphosphate (dCTP) used to label the oligonucleotide probe (Nunez *et al.* 1989), are added to about 650 mg of tissue in microcentrifuge tubes. Any non-volatile source of radioactivity that produces a homogeneous distribution when mixed with tissue paste could be used. In order to obtain a suitable range of optical density values in the resulting autoradiographic image when apposed to radiation-sensitive film, a minimum of about eight standards with different activity levels should be prepared by constructing a dilution curve as follows: 1, 0.66, 0.50, 0.33, 0.25, 0.17, 0.12, 0.06. The concentration of radioactivity selected for the initial solution will depend on the anticipated length of exposure. This will be influenced mainly by the energy of the radionuclide and the specific activity of the probe. Exposure times range from a few hours for ^{32}P, several days for ^{35}S, 1–2 weeks for ^{125}I, and up to 3 months for ^{3}H. Other factors include the abundance of mRNA, the thickness of the tissue section, and the sensitivity of the film used. For example, in order to produce standards 10 μm thick for the quantification of ^{35}S oligonucleotides within auto-radiographic images produced by exposing cardiovascular or neuronal tissue to Hyperfilm βmax (Amersham International) or Kodak X-Omat XS1 for 1–14 days, the initial solution in the dilution series contains about 10 μCi of radioactivity, which is added to 650 mg of tissue.

Tubes are vortexed for 10 min before the tissue is snap frozen in liquid nitrogen. Sections are cut on a cryostat (for example, Bright rotary microtome, Brights, Huntingdon, Cambridge) to the same thickness as the experimental tissue, thaw-mounted onto gelatin-subbed slides and allowed to dry. Representative sections are collected during the cutting process and the average amount of radioactivity associated with each standard is measured in dpm after making appropriate corrections for background and efficiency of counting.

Potential sources of error

A wide variation (greater than 5 per cent) in the radioactive content of replicate sections may indicate uneven cutting. This can be reduced by using a high-quality cryostat and ensuring a reproducible slow cutting action with hand-operated equipment or by using a motor-driven microtome.

Preparation of commercial polymer standards

Commercially prepared standards are available for some of the isotopes which have a short half-life such as ^{35}S (90 d) and ^{125}I (60 d), as well as isotopes with longer half-lives such as ^{3}H. An advantage of these standards is that the dimensions of the microscale are suitable for mounting on microscope slides together with sections of tissue for the quantification of silver grains. Polymer standards from Amersham International consist of layers containing radioactivity incorporated at the molecular level in a methacrylate copolymer, separated by inert coloured layers. They are available in pre-cut strips, or as a multilayer block.

Cutting polymer standards

Multilayer blocks are mounted vertically using a foil clamp onto a standard rotary microtome (for example, a Reichert-Jung 1150 Autocut). Strips are cut at room temperature to the same thickness as the cryostat sections using either steel or glass knives. The cut strips are expanded on water at 60 °C and brushed flat onto gelatin-subbed slides to remove creases. Representative sections are subdivided into the individual activity levels and counted as previously described.

8

mRNA *in situ* hybridization and the study of development

DAVID G. WILKINSON

Introduction

During embryonic development, a single cell—the fertilized egg—gives rise to a number of tissues, each comprised of many cell types organized in a characteristic spatial arrangement. It is convenient to distinguish three types of processes that occur during development, although these are often closely related and may even have some mechanisms in common. First, the generation of body morphology is accomplished through morphogenetic movements, such as the transient migration of certain cell populations and the folding and fusion of epithelial sheets. These processes involve changes in cell behaviour; for example, in their interactions with other cells and with extracellular matrix, and this is presumably mediated in part by cell-surface molecules. Second, cells become progressively more restricted in their potential fate (that is, the spectrum of cell types that they are capable of forming) such that ultimately they are committed to differentiating into a specific cell type. Thus, cell differentiation is preceded by cell determination. Third, differentiated cells (and organs) arise with a defined spatial relationship to each other, a phenomenon known as pattern formation. It is evident that all of these processes must involve the temporal and spatial control of gene expression. There are two major ways of studying the spatial distribution of gene products: to detect the protein through the use of antibodies or to detect mRNA by *in situ* hybridization. A major advantage of the latter technique is the ease with which specific probes can be generated from genomic or cDNA clones, or by the synthesis of oligo-nucleotides. This is a particularly important consideration given the plethora of genes with potential roles in development that are being isolated and characterized, and because it can be time-consuming and difficult to raise specific antibodies. It is for this reason that *in situ* hybridization is, and will continue to be, an important technique in the study of development. In this article, the ways in which *in situ* hybridization has been applied to developmental problems will be illustrated with examples of some recent advances in the understanding of pattern formation in insects and mice.

Technical aspects of *in situ* hybridization

Details of the methods of *in situ* hybridization are to be found in other chapters in this volume. Certain technical aspects of *in situ* hybridization are of particular importance for studies of development, and these will be briefly discussed below.

Choice of probe

The first choice to be made is whether to use DNA or RNA probes. Although single-stranded or double-stranded DNA probes were used in a number of early studies (Akam 1983), RNA probes are now the method of choice owing to their ease of preparation, sensitivity, and ability to use ribonuclease to reduce background. The latter feature is a major advantage, since a common limitation of *in situ* hybridization is the signal-to-noise ratio. However, DNA oligonucleotides may be useful in certain situations, since it is relatively easy and convenient to synthesize transcript-specific probes that may otherwise be difficult to obtain; an application that may find increasing use is the detection of specific transcripts of genes that are alternatively spliced in different tissues or regions of the embryo. It is particularly important to ascertain the specificity of hybridization of probe to cellular RNA, since many genes with developmental roles are members of multigene families (Dressler and Gruss 1988; Mercola and Stiles 1988). Clearly, probes that cross-hybridize to the transcripts of several different genes may give misleading results. Generally, probes derived from untranslated regions, but lacking repetitive elements (and the poly-A tail) are likely to be specific, and this can be ascertained by hybridization to appropriate Northern and genomic Southern blots. The stringency of washing can be altered to retain only specific hybrids, and it is possible to wash tissue sections at high stringency, for example, 0.3 M NaCl, 50% formamide at 65 °C, without significant loss of signal (Angerer *et al.* 1985; Wilkinson *et al.* 1987*b*). As a further test for the possibility of unwanted cross-hybridization, several non-overlapping regions of a gene can be used as probes and the patterns of hybridization compared.

Resolution

It is often desirable to obtain maximal resolution of the signal over the developing tissue, particularly in early-stage embryos where distinctive cells and tissues are in close proximity. Although the use of tritiated probes allows the sharpest resolution, ^{35}S-labelled probes are most frequently used because they give more rapid results at a sufficiently good resolution for many purposes. Some laboratories have even achieved good results using ^{32}P-labelled probes, despite the poor resolution that theory predicts. At present, non-radioactive probes have not found widespread use in developmental studies, despite a number of potential advantages; their sensitivity has not yet been shown to match that of radioactive probes. A factor that indirectly affects resolution is the quality of the tissue histology. Many laboratories have used paraffin-embedded embryos since they give excellent preservation of tissue structure and allow reliable collection of all serial sections. The latter feature is particularly welcome when one wishes to analyse the distribution of transcripts throughout an entire early-stage embryo, or compare the expression of different RNAs using adjacent sections. A related advantage, is that it is relatively easy to orientate embryos in wax to accurately obtain specific planes of section, and this can be crucial for the correct interpretation of results. However, some laboratories have obtained good results with later-stage embryos using frozen sections cut on a cryostat, and this technique may allow slightly greater sensitivity of detection of RNA.

Sensitivity and noise

The most crucial aspects of *in situ* hybridization are also the most frequent sources of problems: sensitivity and non-specific background. Different developmental systems have different problems associated with them; for example, the presence of impermeable membranes or large amounts of yolk. For this reason, various system-specific recipes have evolved and the reader is referred to some original references for details of protocols for sea urchins, *Drosophila, Xenopus* and mice (Cox *et al.* 1984; Ingham *et al.* 1985; Kintner and Melton, 1987; Wilkinson *et al.* 1987*a,b*). Most of these protocols are based on the excellent study of Cox *et al.* (1984), who analysed many of the fundamental aspects of the technique. Cross-linking fixatives, such as paraformaldehyde, have found general favour in many different systems as they provide good retention of RNA while allowing access of probe. As discussed in other chapters, the pretreatment of sections with proteinase further increases sensitivity, as does the reduction of probe length to ~100 bp average. Although it is difficult to determine the limits of sensitivity with confidence, a reasonable estimate is that around 20 transcripts per cell can be detected. The most common limiting factor in detecting low levels of transcripts is the extent of non-specific binding of probe to cellular components. There have been a number of reports of this being a particular problem in certain tissues, for example in testes (Haffner and Willison 1987), and we have often observed a threefold higher background over developing liver in the mouse. It is, of course, necessary to be sure that 'noise' is not due to the widespread low-level expression of the RNA in question. In some cases, this can be examined by, for example, RNase protection assays of RNA in different tissues, but this is not feasible for small early-stage embryos. Although the presence of signal after hybridization with control probes, such as sense-strand RNA, may be indicative of non-specific binding, it is difficult to be certain of this given the variation in noise found with different probes (see below). A good test of whether signals are specific is to compete with an excess of unlabelled antisense RNA identical to the probe: a 20-fold excess drastically reduces specific signals, but has little effect on non-specific binding (unpublished observations). If noise is found to be a problem, then a number of possible cures can be tried. High-stringency washing often reduces background considerably, as does the use of short (< 300 bp) rather than long inserts for the generation of probes. In addition, probes corresponding to different regions of a transcript can give quite different backgrounds. Thus, it may be worthwhile to test several regions of the gene to identify sequences that yield the probe with the lowest non-specific signal (as well as to test for possible cross-hybridization—see above).

In situ hybridization and studies of development

Different developing systems offer distinctive technical advantages for the study of particular aspects of development, and this is reflected by the diverse ways in which *in situ* hybridization has been used. Here, the strengths and limitations of the technique will be illustrated by examples of the use of *in situ* hybridization in the analysis of several fundamental aspects of animal development.

The expression of RNA and protein

The pattern of expression of RNA is generally assumed to correspond to the distribution of the encoded protein. While this assumption may frequently be valid, there are some clear cases where the patterns of protein and RNA differ. The expression of several maternal RNAs in the *Drosophila* embryo provide good examples of this phenomenon. Bicoid is a maternal effect gene whose product is required for the correct development of anterior structures in the *Drosophila* embryo. Analysis of the distribution of bicoid RNA by *in situ* hybridization revealed its localization at the anterior pole of the egg and early embryo (Berleth *et al.* 1988). This pattern of expression is consistent with the developmental role of bicoid and suggests the existence of mechanisms that anchor the RNA to a specific region of the cytoplasm. However, analysis of the accumulation of bicoid protein revealed two significant features of bicoid expression that were not detectable from analysis of the RNA alone. First, bicoid protein does not accumulate until after 1–2 hours of development, and this suggests the existence of translational control ·(Driever and Nusslein-Volhard 1988a). Indeed, the translational control of maternal RNAs seems to be a common feature in many organisms. Second, although bicoid RNA is restricted to the anterior 20 per cent of the early embryo (which is syncytial), bicoid protein forms an exponential gradient that encompasses the anterior 70 per cent of the embryo (Driever and Nusslein-Volhard 1988a). This gradient seems likely to be formed by the diffusion and turnover of the protein, and studies of mutants suggest that it is crucial for the developmental function of the bicoid gene (Driever and Nusslein-Volhard 1988b). An even more striking example of disparity between RNA and protein distributions is provided by analysis of the dorsal gene. Dorsal RNA is uniformly distributed in the Drosophila egg and early embryo, and, like bicoid, its translation is initiated after several hours of development (Steward *et al.* 1988). However, dorsal protein accumulates only in ventral regions of the embryo, a pattern that is consistent with the involvement of this gene in the formation of ventral tissues. This spatially restricted pattern of protein accumulation may be due to differential translational control or its rapid turnover only in dorsal regions of the embryo. These *Drosophila* genes encode nuclear proteins, but in the case of secreted proteins it is perhaps more obvious that their distribution may differ from the sites of RNA accumulation. Immunocytochemical analysis can provide clues regarding the function and possible targets of these proteins that could not be deduced from analysis of the RNA alone. Thus, it is frequently advantageous to examine both RNA and protein accumulation since this information can contribute significantly to the understanding of their developmental role.

RNA as a marker of cell phenotype

The expression of a specific RNA can be regarded as a marker of cell phenotype that can be detected in the absence of any morphological changes. There are now very many examples of cell-type specific RNAs being detected long before overt morphological differentiation, and the use of *in situ* hybridization to detect such RNAs has a number of potential uses. If it is assumed that the RNA is a lineage marker, that is, cells expressing the RNA at an early stage are progenitors of those expressing at later stages (although this is difficult to prove), then the site of origin and movement of specific cells can be followed by the observation of patterns of expression at different times of development. In addition, it may also be possible to analyse the factors that influence cell differentiation in systems where terminal differentiation or morphogenesis is blocked (for example during *in vitro* culture of tissues). *In situ* hybridization has also been used to analyse the expression of genes that are involved in pattern formation in *Drosophila*. The cell phenotype defined by the expression of these genes in the early Drosophila embryo is not associated with cell-type differentiation, but rather with 'cell-states' that lead to the establishment of body pattern (for example, see Martinez-Arias *et al.* 1988). A crucial technique in these studies has been the use of genetics, and this is described in more detail in the next section.

Analysis of development by genetics and *in situ* hybridization

In situ hybridization has been most powerfully applied in developmental studies in combination with the use of genetics to study pattern formation in *Drosophila*. In part, this is because interpretation of the significance of patterns of specific gene expression requires definition of the developmental function of the gene, and the phenotype of mutants is the most incisive test of this. The saturation screening for mutants defective in various aspects of early development has led to the cloning of many of the corresponding genes, and the best-characterized of these are involved in segmentation (the segmentation genes) and defining segment identity (the homeotic genes). *In situ* hybridization analysis has shown that the expression pattern of many of these genes is (largely) as might be predicted from the phenotype of the mutant embryo; the structures deleted or aberrant in the mutant are also the sites of normal expression of the gene. Thus, for example, mutations in the segmentation gene fushi tarazu lead to the deletion of alternate segments, and it is these segments that normally express fushi tarazu RNA and protein (Hafen *et al.* 1984; Carroll and Scott 1985). However, in some cases the effect of a specific mutation can extend beyond the site of gene expression, owing to effects on intercellular interactions (see above). This phenomenon occurs for the wingless gene, which is expressed in a narrow domain within each segment (Baker 1987) (Fig. 8.1(A)), yet wingless mutants have a substantially broader region of each segment deleted. A simple explanation for this is that the wingless protein is secreted and is required for the appropriate development of neighbouring cells; indirect evidence that suggests *int-1* protein is secreted supports this concept (Papkoff *et al.* 1987).

The analysis of developmental mutants and the expression patterns of the corresponding genes suggests that the establishment of body pattern requires the correct spatial control of RNA accumulation. Thus, elucidation of how this control is achieved is of crucial importance for the understanding of development. The use of *in situ* hybridization to study the effects of a specific mutation on the expression of other genes has led to understanding of how the spatial patterns of transcript accumulation are established (reviewed by Ingham 1988). An example of the approach used is shown in Fig. 8.1(A, B). As mentioned above, wingless RNA is expressed in a narrow domain within each segment (Fig. 8.1(A)). However, in an embryo mutant for the patched gene, the domain of wingless expression broadens (Martinez-Arias *et al.* 1988) (Fig. 8.1(B)). In view of the normal expression of the patched gene in the cells adjacent to those expressing wingless, this suggests that the patched gene product serves to repress wingless gene expression. By similar analyses it has been shown that the wingless gene regulates the expression of the engrailed gene in adjacent

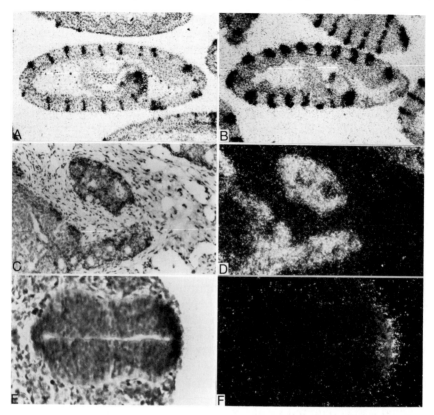

Fig. 8.1. Expression patterns of the *Drosophila* wingless gene and its mouse homologue, the proto-oncogene *int-1*: (A) wingless expression in wild-type *Drosophila* embryo; (B) wingless expression in patched mutant embryo; (C, D) *int-1* expression in mouse mammary tumour; (E, F) *int-1* expression in neural tube of 10.5-day mouse embryo. See Martinez-Arias *et al.* (1988) and Wilkinson *et al.* (1987*a,b*) for further details. ((A) and (B) reproduced from Martinez-Arias *et al.* (1988) with permission.)

cells, consistent with the former encoding a putative secreted protein (Martinez-Arias *et al.* 1988; DiNardo *et al.* 1988). Through the systematic use of this genetic approach, there is now good evidence for a cascade of interactions between genes that regulate each other's expression, either positively or negatively, and sometimes in a combinatorial manner (Ingham 1988). These interactions lead to the spatial control of gene expression which underlies the patterning of the early embryo.

In situ hybridization and vertebrate development

The elucidation of developmental mechanisms in vertebrates is hampered by the difficulty of systematic genetic analysis. In particular, it is not possible to deduce the developmental function of genes by the analysis of mutants. Nevertheless, a number of genes have been identified that seem likely to have important roles in vertebrate development. *In situ* hybridization is an invaluable technique for the characterization of these genes, since the pattern of expression can, together with other information, provide clues as to the role of the gene under study.

An increasingly popular approach towards understanding vertebrate development is to take advantage of the evolutionary conservation of genes that control pattern formation in *Drosophila*. Analysis of the expression patterns of the vertebrate counterparts of these genes suggests that at least some of them may have analogous roles in these diverse developmental systems. This is best illustrated by the studies of the first genes to be identified through this approach: the homeobox genes. The homeobox is a DNA sequence conserved between many genes involved in *Drosophila* embryogenesis (Gehring 1987), and encodes a protein domain that seems to mediate sequence-specific binding to DNA (Desplan *et al.* 1985). It is believed that the homeobox genes encode transcription factors that regulate each other's expression, and are an important component in the network of gene interactions mentioned above. Homeobox genes have been identified in vertebrates by cross-hybridization with *Drosophila* homeobox probes (McGinnis *et al.* 1984). The best characterized of these vertebrate homologues (the *Hox* genes) are a family of more than 20 genes that are related to the *Drosophila* Antennepedia-like homeotic genes. Since the latter specify segment phenotype, and this differs according to the position along the anterior–posterior axis, it was anticipated that *Hox* genes may similarly control region-specific patterns of development. Analysis of the patterns of expression of the *Hox* genes suggests that this may indeed be the case, since all of the *Hox* genes share one striking feature: expression in a restricted anterior–posterior region of the central nervous system (reviewed by Holland and Hogan 1988). Comparison of the expression patterns of different *Hox* genes on adjacent sections of mouse embryos shows that they have different domains of expression in this tissue. In addition, different *Hox* genes have different patterns of expression within other embryonic tissues. This region-specific expression is reminiscent of their *Drosophila* homologues, and suggests that *Hox* genes may specify position along the body axis. Test of this will require genetic manipulation of *Hox* gene expression, for example by mutation or inappropriate expression. While a strong case can be made for the function of *Hox* genes based on their expression patterns, perhaps a more typical situation is illustrated by analysis of the vertebrate homologue of a different *Drosophila* gene. This homologue was first identified as a proto-oncogene, *int-1*, that is implicated in mammary tumour formation in the mouse (Nusse and Varmus 1982). Expression of this gene in epithelial cells of a mammary tumour is shown in Fig. 8.1(C, D). Subsequently, it was found that the *Drosophila* homologue of *int-1* is the segmentation gene wingless (Rijsewijk *et al.* 1987) (see Fig. 8.1(A)). Expression of *int-1* occurs in a restricted population of cells in the developing central nervous system of the mouse (Fig. 8.1(E, F)) Wilkinson *et al.* 1987*b*). However, the role of wingless in *Drosophila* development offers few clues as to the function of *int-1* in nervous-system development beyond the idea that it encodes a secreted protein involved in cell–cell interactions. Thus, speculations regarding *int-1* function are largely based on its pattern of expression, and are not aided by direct analogy with the function of its homologue.

An ever-increasing number of genes that are expressed in the vertebrate embryo are being identified by many other approaches. Some of these genes are related to previously known genes (growth factors, receptors, transcription factors, etc.), and are presumed to have roles that are analogous to those of their relatives. Other genes have been isolated on the basis of screens (proto-oncogenes, growth factor-regulated genes, etc.) that are often intended to identify genes with certain biological functions. The analysis of expression patterns is frequently used to confirm expectations, or to infer the developmental role of these genes. This approach is clearly of great value, but may also have some pitfalls. The pattern of expression of a gene can be used as evidence to argue against certain preconceptions—for example, the detection of proto-oncogene transcripts in post-mitotic neurones strongly suggests roles other than the control of cell proliferation, at least in this tissue. In addition, patterns of expression can suggest specific hypotheses, and, at the very least, these should guide the design of manipulative experiments to test function. However, there are some potential limitations to these interpretations. Since many cell type-restricted differentiation products accumulate at early stages of development, their pattern of expression may be similar to those of genes involved in cell-type specification. Thus, independent evidence must be sought before proposing that the gene under study is involved in the latter process. The fact that translational control may lead to a disparity between RNA and protein accumulation (see above) should also be grounds for caution in interpreting RNA expression patterns. On the other hand, the failure to detect a specific mRNA by *in situ* hybridization does not necessarily mean that biologically significant amounts of the corresponding protein are not present, especially if insensitive methods are being used. A final, and more pessimistic caveat is based on the observation that certain *Drosophila* segmentation genes are expressed in cells whose phenotype is apparently unaffected by mutation of the gene, and thus patterns of expression may in part reflect mechanisms of gene control rather than sites of gene action. Similar conclusions have been reached from studies showing diverse species-specific patterns of expression of alcohol dehydrogenase, and the apparent absence of phenotypic effect of perturbed patterns of glucose dehydrogenase expression in *Drosophila* (Cavener 1987; Dickinson 1988).

Prospects

It is highly likely that *in situ* hybridization will continue to be an important technique in studies of developmental mechanisms. The establishment of methods that allow the genetic manipulation of vertebrates will allow the analysis of gene expression patterns in developmental mutants, an approach that has been powerfully applied to the *Drosophila* embryo. Another approach towards understanding the developmental significance of gene expression patterns is to compare these in different species that have similar modes of development. The assumption behind this approach is that only sites of expression that are important for development are likely to be conserved between species, for example between mice and chicks. In some cases (for example, *Hox* genes; A. Graham, personal communication), cross-species hybridization of probes is possible and thus comparison of expression patterns can be carried out very rapidly. Finally, the establishment of *in situ* hybridization as a routine technique makes it feasible to screen for genes with particular types of developmental expression patterns, and this may allow a more systematic and focused approach towards analysing specific problems in development.

Acknowledgements

I thank Dr Philip Ingham for generously supplying the photographs of wingless expression in *Drosophila* embryos.

References

Akam, M. E. (1983). The location of Ultrabithorax transcripts in *Drosophila* tissue sections. *EMBO J.*, **2**, 2075–84.

Angerer, L., DeLeon, D., Cox, K., Maxson, R., Kedes, L., Kaumeyer, J., Weinberg, E., and Angerer, R. (1985). Simultaneous expression of early and late histone messenger RNAs in individual cells during development of the sea urchin embryo. *Develop. Biol.*, **112**, 157–64.

Baker, N. E. (1987). Molecular cloning of sequences from *wingless*, a segment polarity gene in *Drosophila*: the spatial distribution of a transcript in embryos. *EMBO J.*, **6**, 1765–73.

Berleth, T., Burri, M., Thoma, G., Bopp, D., Richstein, S., Frigerio, G., Noll, M., and Nusslein-Volhard, C. (1988). The role of localisation of bicoid RNA in organising the anterior pattern of the *Drosophila* embryo. *EMBO J.*, **7**, 1749–56.

Carroll, S. B. and Scott, M. P. (1985). Localisation of the fushi tarazu protein during *Drosophila* embryogenesis. *Cell*, **43**, 47–57.

Cavener, D. R. (1987). Combinatorial control of structural genes in *Drosophila*: solutions that work for the animal. *Bioessays*, **7**, 103–7.

Cox, K. H., DeLeon, D. V., Angerer, L. M., and Angerer, R. C. (1984). Detection of mRNAs in sea urchin embryos by *in situ* hybridisation using asymmetric RNA probes. *Develop. Biol.*, **101**, 485–502.

Desplan, C., Theis, J., and O'Farrell, P. (1985). The *Drosophila* developmental gene engrailed encodes a sequence-specific DNA binding activity. *Nature*, **318**, 630–5.

Dickinson, W. J. (1988). On the architecture of regulatory systems: evolutionary insights and implications. *Bioessays*, **8**, 204–8.

DiNardo, S., Shear, E., Heemskerk-Jongens, J., Kassis, J., and O'Farrell, P. (1988). Two-tiered regulation of spatially patterned engrailed gene expression during *Drosophila* embryogenesis. *Nature*, **332**, 604–9.

Dressler, G. R. and Gruss, P. (1988). Do multigene families regulate vertebrate development? *Trends Genet.*, **4**, 214–19.

Driever, W. and Nusslein-Volhard, C. (1988*a*). A gradient of bicoid protein in *Drosophila* embryos. *Cell*, **54**, 83–93.

Driever, W. and Nusslein-Volhard, C. (1988*b*). The bicoid protein determines position in the *Drosophila* embryo in a concentration-dependent manner. *Cell*, **54**, 95–104.

Gehring, W. J. (1987). Homeobox genes in the study of development. *Science*, **236**, 1245–52.

Hafen, E., Kuroiwa, A., and Gehring, W. J. (1984). Spatial distribution of transcripts from the segmentation gene fushi tarazu during *Drosophila* development. *Cell*, **37**, 833–41.

Haffner, R. and Willison, K. (1987). *In situ* hybridisation to messenger RNA in tissue sections. In *Mammalian development* (ed. M. Monk), pp. 199–215.

Holland, P. W. H. and Hogan, B. L. M. (1988). Expression of homeobox genes during mouse development: a review. *Genes Develop.*, **2**, 773–82.

Ingham, P. W. (1988). The molecular genetics of embryonic pattern formation in *Drosophila*. *Nature*, **335**, 25–33.

Ingham, P. W., Howard, K. R., and Ish-Horowicz, D. (1985). Transcription pattern of the *Drosophila* segmentation gene hairy. *Nature*, **318**, 439–45.

Kintner, C. R. and Melton, D. A. (1987). Expression of Xenopus N-CAM RNA in ectoderm is an early response to neural induction. *Development*, **99**, 311–25.

Martinez-Arias, A., Baker, N. E., and Ingham, P. W. (1988). Role of segment polarity genes in the definition and maintenance of cell states in the *Drosophila* embryo. *Development*, **103**, 157–70.

Mercola, M. and Stiles, C. D. (1988). Growth factor superfamilies and mammalian embryogenesis. *Development*, **102**, 451–60.

McGinnis, W., Barber, R. L., Wirz, J., Kuroiwa, A., and Gehring, W. J. (1984). A homologous protein-coding sequence in *Drosophila* homeotic genes and its conservation in other metazoans. *Cell*, **37**, 403–9.

Nusse, R. and Varmus, H. (1982). Many tumors induced by the mouse mammary tumor virus contain a provirus integrated in the same region of the host genome. *Cell*, **31**, 99–109.

Papkoff, J., Brown, A. T., and Varmus, H. (1987). The *int-1* proto-oncogene products are glycoproteins that appear to enter the secretory pathway. *Mol. Cell. Biol.*, **7**, 3978–84.

Rijsewijk, F., Schuermann, M., Wagenaar, E., Parren, P., Weigel, D., and Nusse, R. (1987). The *Drosophila* homolog of the mouse mammary oncogene *int-1* is identical to the segment polarity gene wingless. *Cell*, **50**, 649–57.

Steward, R., Zusman, S. B., Huang, L. H., and Schedl, P. (1988). The dorsal protein is distributed in a gradient in early *Drosophila* embryos. *Cell*, **55**, 487–95.

Wilkinson, D. G., Bailes, J. A., Champion, J. E., and McMahon, A. P. (1987*a*). A molecular analysis of mouse development from 8 to 10 days post coitum detects changes only in embryonic globin expression. *Development*, **99**, 493–500.

Wilkinson, D. G., Bailes, J. A., and McMahon, A. P. (1987*b*). Expression of the proto-oncogene *int-1* is restricted to specific neural cells in the developing mouse embryo. *Cell*, **50**, 79–88.

9

In situ hybridization in virology

C. G. TEO

Introduction

Initial reports of the *in situ* hybridization method (Pardue and Gall 1969; John *et al.* 1969) demonstrated the direct visualization of genes coding for 18S and 28S ribosomal RNA in *Xenopus laevis* oocytes at the cellular, sub-cellular, and chromosomal levels of microscopic resolution. Virologists were not slow to realize how versatile and refined this approach could be because reports soon appeared showing that resolution at these three levels could also be attained in the study of virus-infected systems. Orth *et al.* (1970) demonstrated, in tumours induced by the Shope rabbit papilloma virus (RPV), that RPV genomes could be visualized predominantly in keratinizing cells of the stratum granulosa and cornea but not in prolifer-ating cells of the stratum germinativum and spinosum; moreover, it was found that the extent of viral replication increased as keratinization pro-gressed. Subcellular localization of viral genomes in cytological prepara-tions was achieved by Geukens and May (1974), who provided evidence for the preferential localization of SV40 DNA in the nucleoli of CV1 monkey kidney cells. Finally, viral genes could also be localized to their host chromosomes: proviral DNA of murine sarcoma virus was found to be associated with centromeric heterochromatin of mouse 3T6 cells (Loni and Green 1974). Ensuing improvements in probes and techniques had brought the sensitivity of *in situ* hybridization to such a point that single copies of cellular genes (Gerhard *et al.* 1981; Harper *et al.* 1981) could be chromo-somally assigned with radioactive probes, and non-radioactive probes (see Chapter 10). Soon after, Henderson *et al.* (1983) showed that single viral genes, too, could be mapped on chromosomes.

However, virologists had not merely been emulating those working with cellular genes; they too had contributed significantly to advances in the technique. Brahic and Haase (1978), using sheep choroid plexus cells infected with visna virus, established an exacting, universally applicable *in situ* hybridization protocol that resulted in a highly efficient ^{3}H-cDNA–RNA hybridization process, permitting the quantitative assay of RNA levels in a cell. A similarly sensitive and quantitative method of determining the content of RNA was developed by Harper *et al.* (1986) using cloned

^{35}S-labelled RNA probes specific to human immunodeficiency virus-1 (HIV-1). Several innovative *in situ* hybridization-related approaches have emanated from the laboratory of Ashley Haase in the course of studying virus pathogenesis. These are techniques that: (1) allow sequential images of the anatomical distribution of genes in whole organ or animal sections to be obtained together with the cellular localization of the genes within these sections (Haase *et al.* 1985*a*); (2) enable the simultaneous detection of viral genes and gene products in the same cell (Brahic *et al.* 1984); and (3) permit the identification of cells that contain two different nucleic acid sequences (Haase *et al.* 1985*b*).

Brigati *et al.* (1983), utilizing a range of biotinylated virus-specific DNAs to probe for the presence of virus genomes in actively-infected tissues, showed that non-radioisotopic, immunocytochemical detection of genes by *in situ* hybridization could be performed. Hence the technique could now be directly transplanted not only to diagnostic virology laboratories but routine laboratories of other disciplines as well. Yet another potentially useful approach, initially developed to detect herpes simplex virus DNA in encephalitis tissues, is to perform an immunochemical step after hybridization with probe sequences whose thymidine sequences have been dimerized, using as the first layer antiserum that recognizes dimerized thymidine (Nakane *et al.* 1987). Immunocytochemical detection of hybridized nucleic acid sequences is now so well developed that single copies of viral and cellular genes can be visualized by light microscopy in metaphase chromosomes (Garson *et al.* 1987; Bhatt *et al.* 1988; Lawrence *et al.* 1988; Teo and Griffin 1987) as well as in interphase nuclei (Lawrence *et al.* 1988; Teo and Griffin, 1987). This is illustrated in Fig. 9.1(A–C).

Moreover, the generally low level of background labelling resulting from non-radioisotopic methods also permits sharp discrimination to be made within a single cell of regions containing and those not containing the target genes (Fig. 9.1(D–E)).

The contribution of *in situ* hybridization to animal virology

In situ hybridization is not the only method for determining the presence of a specific virus at the single-cell level, since cytochemical detection of viral antigens can also be used for this purpose. However, there are situations where demonstrating the presence of viral genomes or transcripts may be the only approach. Firstly, antibodies to antigens may not be available. Secondly, artefactual destruction of antigen as a consequence of autolysis, tissue fixation or other processing procedures may occur, as was encountered in a study in which JC virus DNA but not virus antigens was detected in formalin-fixed brain tissues of patients with progressive multifocal

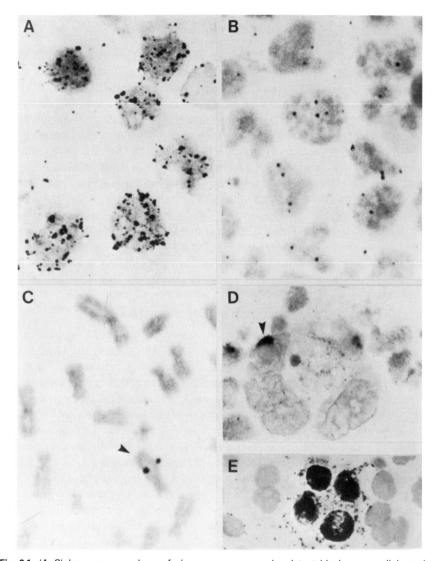

Fig. 9.1. (A–C) Low copy numbers of virus genomes may be detectable by non-radioisotopic methods of *in situ* hybridization. (A) Raji cells following hybridization with an EBV-specific probe. Raji cells have an average EBV genomic copy number of 50–100 per cell. (×1200). (B) Namalwa cells after hybridization with an EBV-specific probe. This cell line has about three EBV genomes per cell. (×1200). (C) Partial metaphase spread of a cell from the IB4 line (which harbours latent EBV) showing EBV genomic integration at 4q2 (arrow) after hybridization with EBV-specific probes. (D–E). Visualization of sites of lytic activation of herpes viruses in multinucleated cells. (×700). (D) Not all nuclei of a multinucleated JJHAN cell infected with human herpesvirus-6 (HHV-6) undergo activation, as suggested by heavy labelling within only one nucleus of this particular cell (arrow) following hybridization with HHV-6-specific probe. (E) In contrast, all nuclei of a multinucleate EBV-immortalized B95-8 cell undergo lytic activation, indicated by the heavy staining of all nuclei within this cell (centre) after hybridization with an EBV-specific probe. Extra-nuclear staining probably represents EBV genomes of encapsidated virions. (In all these figures, a three-layer immunoperoxidase method with gold–silver amplification was used to detect the hybridized probe; the counterstain used was Mayer's haemalum in cell and Giemsa in chromosome preparations).

leukoencephalopathy (PML) (Teo *et al.* 1989). Thirdly, defective viral genomes may be exclusively present. Fourthly, cellular repression of viral genomic replication may occur, as exemplified by the blockade of visna proviral replication in sheep choroid plexus cells *in vivo* (Haase *et al.* 1977). Finally, there may be cellular repression of viral genomic expression, as is the case of visna virus in sheep choroid plexus cells (Brahic *et al.* 1981*a*) and in sheep monocytes (Peluso *et al.* 1985), Theiler's murine encephalitis virus in murine glia (Cash *et al.* 1985), and measles virus in brain cells of patients with subacute sclerosing panencephalitis (SSPE) (Haase *et al.* 1981*a*). The fact that *in situ* hybridization is not subject to these constraints, and can localize the position of viral genomes associated with host chromosomes, have been exploited to study the biology of viruses and the pathogenesis of viral infections and to form the basis for diagnosis and prognosis of viral diseases. The following sections review how the technique has contributed to virology in these respects.

Elucidation of the mode of viral replication

In situ hybridization has been used to study the mode of replication of hepatitis B virus (HBV) in human hepatocytes (Blum *et al.* 1984*b*). On the postulate that if the replication of hepadnavirus DNA proceeds asymmetrically by reverse transcription from an RNA intermediate, there will be an accumulation of large amounts of free single-stranded DNA of one polarity only, termed the minus strand, at the site of replication (Mason *et al.* 1982). (Conceivably, this strand then acts as a template for the viral DNA polymerase to generate the plus strand.) Using strand-specific probes cloned into M13 phage vectors, the cytoplasm of hepatocytes was found to contain minus strand viral DNA in abundance, thus satisfying the prediction of the asymmetric replication model. This is one instance of the technique being successfully applied to verify the replication mode of a virus. In general, however, any virus that produces replication-specific forms of its genome can be analysed in this way.

Definition of the mechanism of herpes virus latency *in vivo*

In situ hybridization has been used to identify the neurons of ganglia as the site of persistence of varicella-zoster virus (Hyman *et al.* 1983; Vafai *et al.* 1988) and herpes simplex virus (HSV) (Galloway *et al.* 1982; Stroop *et al.* 1984). These studies do show some degree of viral RNA expression. The molecular events underlying the latent state of HSV, in particular, have been closely dissected using *in situ* hybridization. Stevens *et al.* (1987) showed that in murine spinal ganglia infected by a non-invasive HSV-1 strain, only RNA transcripts derived from a region encoding an immediate early protein, ICP-0, and not other known immediate early proteins, was present in relative abundance. These transcripts were restricted to the nucleus. Similar results were reported in murine trigeminal ganglia by

Deatley *et al.* (1987) and in murine brain stem neurons by Deatley *et al.* (1988). Unexpectedly, Stevens *et al.* (1987) also found that the RNA was antisense to that encoding ICP-0 mRNA. Similar data were subsequently reported in trigeminal ganglia of rabbits (Rock *et al.* 1987) and of humans (Croen *et al.* 1987; Stevens *et al.* 1988). These studies have led to the hypo-thesis that these nuclear 'anti-ICP-0' transcripts, unique to the latent state, function as natural antisense RNAs that regulates expression of the ICP-0 product normally required for productive HSV replication. A further study (Wechsler *et al.* 1988) has shown that there are at least two such transcripts, that they are produced by alternative splicing, but that they contain two potential open reading frames. A more recent and precise study has, how-ever, indicated that the transcripts are not essential for latent infection but rather, reactivation (Steiner *et al.* 1989).

Identification of vehicles of virus dissemination and transmission

This type of analysis is important in determining which host cells are being used by the virus to spread from one site to another. *In situ* hybridization has been used to elucidate the mode of transmission of cytomegaloviruses (CMV). The demonstration of murine CMV DNA in spermatozoa (Dutko and Oldstone 1979) and embryos of mice (Chantler *et al.* 1979) suggests possible mechanisms for natural horizontal and vertical transmission. The detection of human CMV DNA in peripheral blood leukocytes bearing CD4 and CD8 markers (Schrier *et al.* 1985), using *in situ* hybridization and fluorescence-activated cell sorting, specifically identifies lymphocytes as the vehicles for transmission.

Insight into how JC virus may disseminate from potential reservoirs in the body to the brain, allowing consequential association with PML, has been provided by Houff *et al.* (1988) who identified lymphocytes as the possible vehicle. JC DNA was found in kappa light-chain bearing lympho-cytes in the spleen and bone marrow, as well as in intra- and extra-vascular mononuclear cells in the brain of PML patients. Some of these cells were also positive for viral capsid antigen.

In situ hybridization has also been used to extend the evidence that epithelial cells of the oropharynx and salivary glands are sites of persistent infection by the Epstein–Barr virus (EBV) (Sixbey *et al.* 1984; Wolf *et al.* 1984; Venables *et al.* 1989) (Fig. 9.2). These cells are the source of virus that is shed into the saliva, which therefore acts as the vehicle of trans-mission.

In the case of the persistent infection of sheep by visna virus, the virus specifically infects circulating monocytes, and then only those that are maturing to macrophages (Gendelman *et al.* 1986). The requirement for a stage of differentiation to pre-exist prior to viral replication applies also to tissue macrophages (Gendelman *et al.* 1985), which act as reservoirs of

In situ *hybridization*

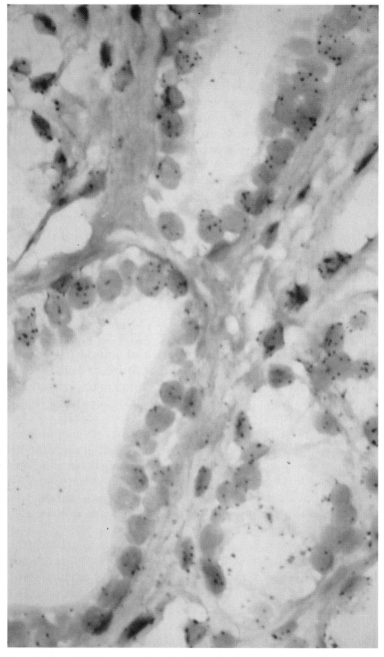

Fig. 9.2. Localization of EBV genomes in epithelial cells of acini and ducts of a salivary gland. Particulate labelling within nuclei results from the hybridization of EBV-specific probe to a labial biopsy frozen section. The counterstain used was Mayer's haemalum (×1000). (Photograph courtesy of Dr. P. J. W. Venables.)

infection. But it has been a puzzle how visna virus spreads continually through the blood, cerebrospinal fluid, and other body fluids of sheep in the presence of neutralizing antibody and immune cells. The finding that visna viral RNA, at least in cerebrospinal fluid monocytes, is restricted from accumulating in these cells, has provided the basis for the 'Trojan horse' mechanism of virus dissemination: the usage by a virus of an infected mobile cell to conceal and convey the viral genome without detection (Peluso *et al.* 1985).

Localization of sites of active viral infection

This category of detection stems from a straightforward application of *in situ* hybridization. The presence of high copy numbers of viral genomes in particular cell types can be taken to indicate productive infection. But to establish firmly this state of productive infection requires the demonstration of the presence in the same cell type of replicative forms of the viral genome, of 'late' viral antigens, and of cytopathy. A classical study of this nature is that of HBV infection of the liver, where the observation of large amounts of viral DNA in hepatocytes within morphologically damaged regions of the liver that were also expressing the surface antigen, provides unequivocal evidence of HBV replication in hepatocytes (Gowans *et al.* 1981). However, less tangible is the claim from a later study, on the sole basis of large amounts of HBV DNA in bile duct epithelium and endothelial and smooth muscle cells of hepatic blood vessels, that these non-hepatocytes are also sites of HBV replication (Blum *et al.* 1983). In this case, the fact that these other cell types could represent sites of sequestration or transit rather than replication was not considered. There are two ways to make this form of discrimination. One is to use strand-specific probes. Jilbert *et al.* (1987) probed for the presence of the single-stranded form of duck hepatitis B virus (DHBV) DNA as evidence for the presence of ongoing replication. Thus, in infected Pekin ducks, DHBV replication was found in hepatocytes and pancreatic cells. Splenic mononuclear cells, where only duplex viral DNA could be detected, was considered to contain only complete virions. The other approach is to examine concurrently the same tissues by blot hybridization methods for replication intermediates. Korba *et al.* (1988) used this approach to examine the extent of tropism of woodchuck hepatitis virus within hepatic and extra-hepatic tissues.

An elegant *in situ* hybridization determination of the site(s) of viral replication is provided by Alexandersen *et al.* (1987), who studied Aleutian disease virus (ADV). Because this parvovirus has a single-stranded DNA genome, it was possible, using strand-specific probes, to distinguish between mink cells harbouring the duplex replicative form of DNA and those merely containing virion DNA. Further evidence of replication was afforded by demonstrating the presence of capsid proteins. In the infected neonatal minks studied, replicative forms of ADV DNA were found predominantly in alveolar type II cells of the lung; DNA in other organs was mainly of the virion type, suggesting virus sequestration in these sites.

In cells undergoing productive infection by human CMV, cytomegaly and intranuclear owl-eye inclusion bodies are well-recognized hallmarks. But when initial *in situ* hybridization studies of disseminated infections were performed, surprisingly, cells within the lungs, heart, liver, spleen, uterus, fallopian tube, breast, kidney, and adrenal glands showed strong positive labelling for CMV DNA (Myerson *et al.* 1984*a,b*). These positively labelled cells were of epithelial, mesenchymal and endothelial origin. They included many that were histologically occult, indicating the increased sensitivity of CMV DNA detection over the examination for cytomegaly or inclusion bodies. Similar results were obtained from later studies (Loning *et al.* 1986; Unger 1986; Keh and Gerber 1988; Raap *et al.* 1988). These do confirm that a very broad range of cell types is permissive of CMV replication and account for the devastating effect of a disseminated infection.

An insight into how a virus can infect the inner ear, subsequently causing deafness by damaging the auditory and vestibular organs, was provided by the visualization of pseudorabies virus DNA in cranial structures of intranasally infected mice (Falser *et al.* 1986). The infected sites were the organ of Corti, various labyrinthine cell types (ganglion, sensory and glial cells), and cells within ganglia of the nucleus cochlearis and cerebellum. The results strikingly show how a neurotropic virus can ascend and infect neural structures at multiple levels. The methodology in this report was remarkable because of the ingenious use of processing techniques to study bony and cartilaginous structures.

Localization of sites of persistent viral infections

Certain viruses may persist in their host without (or after) an acute infectious phase. The virus during this persistent stage may or may not undergo a full productive cycle. However, even a restricted expression of viral products may interfere with host metabolism or elicit immunopathological responses, thereby leading to disease, often of a chronic form. As is often the case, insight into the pathogenesis of some of these diseases has been provided by the direct visualization of viral genomes or transcripts in cells of affected tissues.

In the virus-associated demyelinating diseases, the detection of JC virus DNA in oligodendrocytes of PML cases (Dorries *et al.* 1979; Aksamit *et al.* 1985; Sharpsack *et al.* 1986) may not be considered remarkable because virus capsid antigens and viral particles have previously been demonstrated in such cells. However, *in situ* hybridization studies of the demyelination phase of Theiler's murine encephalitis have yielded more significant and intriguing results; they suggest that ongoing viral replication is not a prerequisite to demyelination. The first phase of this picornavirus infection is an encephalitis, during which there is evidence of full viral replication in the neurons. During the post-encephalitic demyelinating phase, the virus does not replicate in the neurons or other cell types but its genome persists in low amounts in glial cells of the white matter of the spinal cord (Brahic *et al.* 1981*b*). The virus in these cells is restricted at the level of genomic RNA replication (Cash *et al.* 1985). Notwithstanding this, demyelination occurs (Chamorro *et al.* 1986); perhaps the expression of other early viral products could have led to the eventual damage. An even less-direct mechanism of demyelination is found in Marek's disease of fowl, where DNA of Marek's disease virus has been detected not in Schwann cells but in infiltrating lymphocytes of the peripheral nerves (Ross *et al.* 1981). This implies that demyelination is related to the presence of virus-infected lymphocytes and not directly to an immune reaction to viral products in Schwann cells.

In SSPE, measles virus is known to persist in neurons and glia without undergoing the full virus replication cycle. Several *in situ* hybridization studies performed on SSPE tissues have led to a greater insight into its pathogenesis. One group of studies showed that viral RNA in infected brain cells is located primarily in the cytoplasm and occasionally in the dendritic processes (Haase *et al.* 1981*a,b*), suggesting the possibility of cell-to-cell transmission. Furthermore, lymphocytes in the peripheral blood, perivascular brain infiltrate and appendix of SSPE individuals, while not expressing detectable measles antigens, contain viral RNA (Fournier *et al.* 1985, 1986); this raises the possibility that the brain in SSPE could have been infected by lymphocytes in the Trojan horse manner. Finally, a molecular mechanism behind SSPE was defined by the finding that in the early and late stages of the disease, infected cells contain the RNA strands encoding for both matrix (M) and nucleoprotein (NP) antigens, but only NP and not M antigen is expressed (Haase *et al.* 1985*c*). This suggests that the specific defect is at the level of M protein translation.

An attempt to elucidate the pathogenesis of chronic active hepatitis (CAH) was made using a technique to visualize simultaneously HBV DNA and antigens in the same tissue section (Blum *et al.* 1984*b*). Most hepatocytes that contained replicating viral DNA (i.e. DNA in the cytoplasm) did not express detectable core (HBc) antigen, while, conversely, the cells that did express HBc did not harbour replicating DNA. Since cells expressing the HBc antigen are possible targets for T-cell lysis, the dissociation between expression of replicating viral DNA and HBc antigen in patients with CAH may explain why infection remains persistently active.

In most cases of acquired immunodeficiency disease (AIDS) and AIDS-related complex, it is generally accepted that HIV-1 is causally associated with these states. But how this virus can lead to marked depletion of helper-T lymphocytes and lymph node hyperplasia has not been answered with certainty. A careful *in situ* hybridization study (Harper *et al.* 1986) showed one mechanism that may not be operative: the direct infection of CD4 bearing cells, leading to lysis of these cells, thereby accounting for helper-T lymphopenia, or to proliferation of HIV-1-infected lymphocytes, resulting in lymphadenopathy. This is based on the finding that very low numbers (less than 1 in 10^4–10^5) of circulating and tissue lymphocytes express HIV-1 RNA.

Not all persistent virus infections lead to disease. Several *in situ* hybridization studies have demonstrated the persistence of group C adenovirus genomes in tissues containing human lymphoid cells (Horvath *et al.* 1986; Neumann *et al.* 1987). Such persistence appears to be innocuous even though virus transcripts may be present (Jones *et al.* 1979).

Facilitation of tissue tropism studies

The localization of sites of viral replication and persistence, as described in the foregoing two sections, is one way of studying tissue tropism. But this is highly selective in its approach, in that only small samples from the whole organism can be studied at any one time. Another way to determine the tissue tropism of a virus *in vivo* has been to adopt the combined macroscopic–microscopic hybridization technique discussed earlier to screen for the presence of virus genomes in whole organs or small animals; the localization of HBV genes in chronic hepatitis, visna virus genes in sheep brain, and measles virus in SSPE brain was done in this manner (Haase *et al.* 1985*a*,*b*).

An equally comprehensive approach has been to inoculate susceptible embryos with the virus of interest and then examine whole-embryo sections for positive labelling after hybridization with viral probes. Rous-associated virus was found in this way to exhibit marked tropism for skeletal muscle cells of chicken embryos (Flamant *et al.* 1987). Similarly, bluetongue virus showed tropism for epithelial cells of various organs of chicken embryos (Wang *et al.* 1988).

The simultaneous *in situ* assay for antigens and viral genomes was used, together with electron microscopy, to identify sinusoidal-lining cells of the red pulp as the primary cells involved in acute MCMV infection of the spleen (Mercer *et al.* 1988). This finding is intriguing because it shows how 'unconventional' a cell type can be for preferential viral replication.

Provision of presumptive evidence for the viral causation of chronic diseases of obscure aetiology

In situ hybridization can be used in conjunction with other methods to investigate whether a virus is the cause of a chronic disease. The demonstration of virus genomes or transcripts within diseased tissues provides presumptive evidence of a cause-and-effect relationship between virus and disease. Virus-disease associations of this nature include HIV-1 and AIDS-encephalopathy (Shaw *et al.* 1985), human parvovirus and aplastic anaemia

(Kurtzman *et al.* 1987) and hydrops fetalis (Porter *et al.* 1988), measles virus and Paget's disease (Basle *et al.* 1986), and various viruses and neoplasia.

To determine whether a previously uncharacterized virus is associated with a disease, it is possible to hybridize the diseased tissues under low-stringency conditions to probe for known viruses that may be phylogenetically related to the putative virus. Because the genome of this virus may share some homology with the probe, positive labelling may be revealed. A positive result can only be considered circumstantial and should lead to more definitive studies. This approach was used in conjunction with seroepidemiological studies to establish an association between multiple sclerosis and a virus possibly related to human T-leukaemia/lymphoma virus type 1 (HTLV-1) (Koprowski *et al.* 1985). Nevertheless, performing *in situ* hybridization using heterologous probes under low-stringency conditions is technically difficult and the findings are subject to dispute; the failure of an independent study (Hauser *et al.* 1986) to produce results consistent with those of Koprowski *et al.* probably reflects this.

Establishment of a link between virus infections and carcinogenesis

In situ hybridization has been of great value in determining the relationship between certain DNA oncogenic viruses and their transformed hosts. At the most basic level, the direct demonstration of virus genomes in neoplastic cells supports a close association between the virus and the neoplasia. Such an association has apparently been found between SPV and rabbit papilloma (Orth *et al.* 1970), EBV and nasopharyngeal carcinoma (NPC) (Wolf *et al.* 1973) and certain lymphomas (Jones *et al.* 1988), HSV-2 and cervical carcinoma (McDougall *et al.* 1980, Maitland *et al.* 1981), and human papilloma viruses (HPV) and genital condylomata (Beckmann *et al.* 1985) and various grades of cervical dysplasia (Crum *et al.* 1986; Gupta *et al.* 1985).

The consistent presence in specific tumours of certain subtypes of an oncogenic virus lends further support to the tumour–virus association. Numerous *in situ* hybridization studies have provided evidence to suggest a possible relationship between HPV types 16 and 18 and invasive cervical carcinoma by detecting, in a variable proportion of the carcinoma cells, HPV DNA related to these genotypes (Grussendorf-Conen *et al.* 1985; Gupta *et al.* 1985; Ostrow *et al.* 1987; Schneider *et al.* 1987; Collins *et al.* 1988) and mRNA (Stoler and Broker 1987). In these studies, however, the distribution of HPV was distinct from that observed in condylomata and the less advanced grades of cervical intra-epithelial neoplasia, in that there was no increase in labelling extending from the lower epithelial layer to the surface. It is not possible to ascertain any cause-and-effect relationship for this pattern, since either a change in the state of the HPV genome (e.g. integration into host chromosomes) may impede differentiation, or non-differentiated cells may inhibit viral replication.

Causal inferences, however, are difficult to make even if genomes of oncogenic viruses are present in tumours. It is possible that the neoplastic state of the host cell may in some ways be conducive to the persistence or replication of viral genomes without regard to oncogenicity. Conversely, the absence of virus in a tumour does not indicate that the tumour was not caused by the virus, since the virus may be essential for the initiation of the tumour but not its maintenance, as exemplified in the case of papillomas of the alimentary tract of calves caused by bovine papilloma virus (BPV 4) (Jarret 1985).

Notwithstanding this, possible mechanisms of viral carcinogenesis may be inferred by adopting another *in situ* hybridization approach: the analysis of viral integration sites within chromosomes of transformed cells lines (see Table 9.1). This has yielded two probable mechanisms of viral carcinogenesis. The first is the initiation of transformation events following integration near a fragile site, as sugggested by the coincidence between a viral integration site and a fragile site (Simon *et al.* 1985; Popescu *et al.* 1987*a,b*) (see Table 9.1). This coincidence may nevertheless be merely fortuitous. Furthermore, a possible clastogenic action by a virus need not be followed by carcinogenesis. The second, and possibly more plausible, mechanism, is the initiation of events following viral integration at or near sites of proto-oncogenes. Molecular genetic studies of certain retroviruses have demonstrated that integration of viral DNA adjacent to a proto-oncogene can lead to increased activation of the latter. In this regard, the correspondence of HPV 18 integration sites to several proto-oncogene locations in HeLa chromosomes (Popescu *et al.* 1987*a*) may be significant (see Table 9.1).

Certainly, it cannot be ascertained from such studies that integration precedes transformation. For example, the study of the role of EBV genome integration in immortalized lymphoid cell lines does not appear to support any direct link between immortalization and integration. In certain lines, preferential integration in 'hot spots' in the host chromosome does occur, but in others, apparently random integration is found (Table 9.1). Moreover, as shown in the case of transformation of mouse cells by extra-chromosomal BPV 1 genome, transformation need not require integration of viral DNA into the host genome (Law *et al.* 1981). Viral integration may therefore be epiphenomenal to immortalization or transformation.

Development of viral infectivity assays

An adaptation of *in situ* hybridization was used to devise a general infectivity assay for retroviruses. After a monolayer culture of susceptible

Table 9.1. Chromosomal localization by *in situ* hybridization of genomes of DNA viruses

Virus	Source of chromosomes*	Location	Reference
EBV	Raji cells	(Random distribution)	zur Hausen *et al.* 1972
	Namalwa cells[1]	1p35	Henderson *et al.* 1983
	IB4 cells[1]	4q25	
	B-lymphoblastoid cells from individuals with Bloom's syndrome	1p31, 1q31, 4q22 24, 5q21, 13q21, 14q21	Shiraishi *et al.* 1985
	Cells from several Burkitt lymphoma lines	(Random distribution)	Harris *et al.* 1985
	P3HR-1 cells[1]	1q31, 5q2, 13q2, 21q1	Trescol-Biemont *et al.* 1987
	Jijoye cells[1]	13q2	
	Raji cells[1]	1q31, 3q13, 7q21, 7q31, 8q23 21q1	
	IB4 cells[1]	4q2	Teo and Griffin 1987
	AW-Ramos[1]	(Random distribution)	
HSV-1	Murine L cells	Centromeric heterochromatin of 2 translocated chromosomes	Henderson *et al.* 1981
MDV	MDCC[2]	2,4	Hirai *et al.* 1986
Ad12	HEL cells	(Random distribution)	McDougall *et al.* 1972
	T637[3]	2,7,11,12,13,15	Vogel *et al.* 1986
	A2497-2[3]	7,20	
FAV	3Y1[4]	(Random distribution)	Ishibashi *et al.* 1987
SV40	53-87(3) cl.10 & 53-87(1) cl.21 cells[5]	7q31[a]	Rabin *et al.* 1984
HBV	Hep 3B 2-1/7 cells	12q13-q14[a]	Simon *et al.* 1985
HPV	HeLa cells	8q23-q24[a,b], 9q31-q34[c], 5p,11p13,22q12-q13[d]	Popescu *et al.* 1987a
	SW756	12q13[a]	Popescu *et al.* 1987b

* Human unless otherwise indicated.
[1] From human B-lymphoblastoid lines; [2] from chicken T-lymphoblastoid lines; [3] from transformed human–mouse hybrid somatic lines; [4] clonal line of 3T3-type normal rat embryonic cell; [5] from transformed hamster lines.
[a] Coincides with the location of a fragile site; [b] coincides with the location of myc proto-oncogene; [c] coincides with the location of abl proto-oncogene; [d] coincides with the location of sis proto-oncogene.
Abbreviations for viruses not mentioned in the text. Ad: adenovirus; FAV: fowl adenovirus.

cells was infected by a retrovirus, it was hybridized *in situ* with virus-specific ^{32}P-labelled DNA probes; foci of infected cells in the monolayer were detected by exposure of the culture to X-ray film (Rein *et al.* 1982). This method was applicable to a broad range of retroviruses tested and had a sensitivity level similar to conventional plaque assays. Its universality can be widened to include other viruses that can permissively infect cells grown in monolayers.

Formation of the basis of viral diagnosis

In an agricultural field setting, a procedure for the diagnosis of bluetongue virus in sheep was developed by inoculating virus from the blood of an infected sheep to embryonating chicken eggs, and after 24 h examining the embryo sections for viral genomes by *in situ* hybridization (Wang *et al.* 1988). Compared to observing haemorrhage in inoculated eggs, which can take 2–3 weeks to become manifest, this method is undoubtedly much more rapid.

With regard to human infections, it is apparent from the foregoing sections that the confirmation of the diagnosis of CMV infections and PML by histopathologists using *in situ* hybridization is potentially possible. This also applies to HSV encephalitis (Burns *et al.* 1986). *In situ* hybridization has also formed the basis of the diagnosis of human enterovirus heart diseases, in particular, myocarditis and cardiomyopathy. By hybridizing endomyocardial biopsy specimens with enterovirus probes, positive labelling provides unequivocal evidence of enteroviruses in myocardial cells (Kandolf *et al.* 1987). As is the case for other viruses, use of non-radioactively labelled probes will facilitate wider acceptance of this approach. Indeed, steps have been taken to develop this (Easton and Eglin 1988).

The presence of virus in tissue samples can also be determined by *in situ* hybridization following inoculation of the infected sample into permissive cell lines that serve to amplify the viral replicative state. Janssen *et al.* (1987) showed that this approach could be potentially useful for the detection of CMV.

For the rapid diagnosis of viral infections, the direct detection by immunofluorescence of viral antigens in samples is an established method. Because reading of the results can be subjective and a simple permanent recording of the findings is not possible, diagnosis by *in situ* hybridization may be a better alternative. In this respect, the study by Gomes *et al.* (1985) on the presence of adenovirus DNA in cells from nasopharyngeal secretions of children is significant. It was shown that the rapid results obtained from *in situ* hybridization with biotinylated probes correlated well with those of immunofluorescence antigen detection in the same samples. Similarly, the rapid detection of HSV DNA in genital samples using a commercially available biotinylated probe was also found to be both rapid and sensitive (Langenberg *et al.* 1988).

While such 'rapid' methods of diagnosis by *in situ* hybridization may match standard immunofluorescence techniques in terms of sensitivity and time taken from tissue processing to the reading of results, they are certainly not 'convenient'. This is so in those protocols with lengthy pre-hybridization and post-hybridization washes, all of which are unneccesary with an immunofluorescence approach. But there are no great theoretical constraints to the automation of the entire process, including the detection steps following hybridization. When this becomes fully developed it will facilitate a more universal application of *in situ* hybridization to routine virus diagnosis. Already, HPV detection in biopsies can be done in 3–4 hours with non-isotopic methods of good sensitivity and resolution (see Chapter 12).

Prognostication of the malignant potential of cervical condylomatous and dysplastic lesions

The successful *in situ* detection of DNA sequences specific to particular HPV types in koilocytes from cervical Papanicolaou-stained smears (Gupta *et al.* 1987) suggests that it is possible to assess the malignant potential of condylomas and dysplastic lesions. This stems from the gathering evidence that benign lesions containing DNA from HPV 16, 18, 31, and 35 run a high risk of progression to malignancy. Individuals with abnormal cervical smears positive for such HPV types may therefore need to be more closely followed up to detect early invasive changes. This will be an instance of *in situ* hybridization being used as a prognostic tool.

Future directions

The next few years will see the automation of *in situ* hybridization in terms of tissue preparation, hybridization, hybrid detection, and data processing. More recent techniques, e.g. *in situ* transcription (Tecott *et al.* 1988) and mRNA phenotyping (Rappollee *et al.* 1988) will supplement existing approaches in answering questions posed by the virus–host interaction. In turn, still more sophisticated techniques may also be expected to be conceived and developed to answer such questions.

Acknowledgements

I thank Professor B. E. Griffin for critically reviewing the manuscript.

References

Aksamit, A. J., Mourrain, P., Sever, J. L., and Major, E. O. (1985). Progressive multi-focal leukoencephalopathy: investigation of three cases using *in situ* hybridization with JC virus biotinylated DNA probe. *Ann. Neurol.*, **18**, 490–6.

Alexandersen, S., Bloom, M. E., Wolfinbarger, L., and Race, R. E. (1987). *In situ* molecular hybridization for detection of Aleutian mink disease parvovirus DNA by using strand-specific probes: identification of target cells for viral replication in cell cultures and in mink kits with virus-induced interstitial pneumonia. *J. Virol.*, **61**, 2407–19.

Basle, M. F., Fournier, J. G., Rozenblatt, S., Rebel, A., and Bouteille, M. (1986). Measles virus RNA detected in Paget's disease bone tissue by *in situ* hybridization. *J. Gen. Virol.*, **67**, 907–13.

Beckmann, A. M., Myerson, D., Daling, J. R., Kiviat, N. B., Fenoglio, C. M., and McDougall, J. K. (1985). Detection and localization of human papillomavirus

DNA in human genital condylomas by *in situ* hybridization with biotinylated probes. *J. Med. Virol.*, **16**, 265–73.

Bhatt, B., Burns, J., Fleming, D., and McGee, J. O'D. (1988). Direct visualization of single copy genes on banded metaphase chromosomes by nonisotopic *in situ* hybridization. *Nucl. Acids Res.*, **16**, 3951–61.

Blum, H. E., Stowring, L., Figus, A., Montgomery, C. K., Haase, A. T. and Vyas, G. N. (1983). Detection of hepatitis B virus DNA in hepatocytes, bile duct epithelium and vascular elements by *in situ* hybridization. *Proc. Natl. Acad. Sci. USA*, **80**, 6685–8.

Blum, H. E., Haase, A. T., Harris, J. D., Walker, D. and Vyas, G. N. (1984*a*). Asymmetric replication of hepatitis B virus DNA in human liver: demonstration of cytoplasmic minus-strand DNA by blot analyses and *in situ* hybridization. *Virology*, **139**, 87–96.

Blum, H. E., Haase, A. T., and Vyas, G. N. (1984*b*). Molecular pathogenesis of hepatitis B virus infection: simultaneous detection of viral DNA and antigens in paraffin-embedded liver sections. *Lancet*, 771–5.

Brahic, M. and Haase, A. T. (1978). Detection of viral sequences of low reiteration frequency by *in situ* hybridization. *Proc. Natl Acad. Sci. USA*, **75**, 6125–9.

Brahic, M., Stowring, L., Ventura, P. and Haase, A. T. (1981*a*). Gene expression in visna virus infection in sheep. *Nature*, **292**, 240–2.

Brahic, M., Stroop, W. G., and Baringer, J. R. (1981*b*). Theiler's virus persists in glial cells during demyelinating disease. *Cell*, **26**, 123–8.

Brahic, M., Haase, A. T., and Cash, E. (1984). Simultaneous *in situ* detection of viral RNA and antigens. *Proc. Natl. Acad. Sci. USA*, **81**, 5445–8.

Brigati, D. J., Myerson, D., Leary, L. L., Spalholz, B., Travis, S. Z., Fong, C. K. Y., Hsiung, G. D., and Ward, D. C. (1983). Detection of viral genomes in cultured cells and paraffin-embedded tissue sections using biotin-labeled hybridization probes. *Virology*, **126**, 32–50.

Broker, T. R. and Botchan, M. (1986). Papillomaviruses: retrospectives and prospectives. In *DNA tumour viruses: control of gene expression and replication (Cancer Cells 4)*, (ed. M. Botchan, T. Grodzicker, and P. A. Sharp), pp. 17–36. Cold Spring Harbor: Cold Spring Harbor Laboratory.

Burns, J., Redfern, D. R. M., Esiri, M. M., and McGee, J. O'D. (1986). Human and viral gene detection in routine paraffin embedded tissue by *in situ* hybridization with biotinylated probes: viral localization in herpes encephalitis. *J. Clin. Pathol.*, **39**, 1066–73.

Cash, E., Chamorro, M., and Brahic, M. (1985). Theiler's virus RNA and protein synthesis in the central nervous system of demyelinating mice. *Virology*, **144**, 290–4.

Chamorro, M., Aubert, C., and Brahic, M. (1986). Demyelination lesions due to Theiler's virus are associated with ongoing central nervous system infection. *J. Virol.*, **57**, 992–7.

Chantler, J. K., Misra, V., and Hudson, J. B. (1979). Vertical transmission of murine cytomegalovirus. *J. Gen. Virol.*, **42**, 621–9.

Collins, J. E., Jenkins, D., and McCance, D. J. (1988). Detection of human papillomavirus DNA sequences by *in situ* DNA-DNA hybridisation in cervical intraepithelial neoplasia and invasive carcinoma: a retrospective study. *J. Clin. Pathol.*, **41**, 289–95.

Croen, K. D., Ostrove, J. M., Dragovic, L. J., Smialek, J. E., and Straus, S. E., (1987). Latent herpes simplex virus in human and trigeminal ganglia: detection of an immediate early gene 'anti-sense' transcript by *in situ* hybridization. *N. Engl. J. Med.*, **317**, 1427–32.

Crum, C. P., Nagai, N., Levine, R. U., and Silverstein, S. (1986). *In situ* hybridization analysis of HPV 16 DNA sequences in early cervical neoplasia. *Am. J. Pathol.* **123**, 174–82.

Deatley, A. M. *et al.* (1987). RNA from an immediate early region of the type 1 herpes simplex virus genome is present in the trigeminal ganglia of latently infected mice. *Proc. Natl Acad. Sci. USA*, **84**, 3204–8.

Deatley, A. M., Spivack, J. G., Lavi, E., and Fraser, N. W. (1988). Latent herpes simplex virus type 1 transcripts in peripheral and central nervous system tissues of mice map to similar regions of the viral genome. *J. Virol.*, **62**, 749–56.

Dorries, K., Johnson, R. T., and ter Meulen, V. (1979). Detection of polyoma virus DNA in PML-brain tissue by (*in situ*) hybridization. *J. Gen. Virol.* **42**, 49–57.

Dutko, F. J., and Oldstone, M. B. A. (1979). Murine cytomegalovirus infects spermatogenic cells. *Proc. Natl. Acad. Sci. USA*, **76**, 2988–91.

Easton, A. J. and Eglin, R. P. (1988). The detection of coxsackie RNA in cardiac tissue by *in situ* hybridization. *J. Gen. Virol.*, **69**, 285–91.

Falser, N., Bandtlow, I., Haus, M., and Wolf, H. (1986). Detection of pseudorabies virus DNA in the inner ear of intranasally infected BALB/c mice with nucleic acid hybridization *in situ*. *J. Virol.*, **57**, 335–9.

Flamant, F., le Guellec, D., Verdier, G., and Nigon, V. M. (1987). Tissue specificity of retrovirus expression in inoculated avian embryos revealed by *in situ* hybridization to whole-body section. *Virology*, **160**, 301–4.

Fournier, J. G., Lenon, P., Bouteille, M., Goutieres, F., and Rosenblatt, S. (1986). Subacute sclerosing panencephalitis: detection of measles virus RNA in appendix lymphoid tissue before clinical signs. *Br. Med. J.*, **293**, 523–4.

Fournier, J. G. *et al.* (1985). Detection of measles virus RNA in lymphocytes from peripheral blood and brain perivascular infiltrates of patients with subacute sclerosing panencephalitis. *N. Engl. J. Med.*, **313**, 910–15.

Galloway, D. A., Fenoglio, C. M., and McDougall, J. K. (1982). Limited transcription of the herpes simplex virus genome when latent in human sensory ganglia. *J. Virol.*, **41**, 686–93.

Garson, J. A., van den Berghe, J. A., and Kemshead, J. T. (1987). Novel non-isotopic *in situ* hybridization technique detects small (1 Kb) unique sequences in routinely G-banded human chromosomes: fine mapping of N-myc and beta-NGF genes. *Nucl. Acids Res.*, **15**, 4761–70.

Gendelman, H. E., Narayan, O., Molineaux, S., Clements, J. E., and Ghobti, Z. (1985). Slow, persistent replication of lentiviruses: role of tissue macrophages and macrophage precursors in bone marrow. *Proc. Natl Acad. Sci. USA*, **82**, 7086–90.

Gendelman, H. E. *et al.* (1986). Tropism of sheep lentiviruses for monocytes: susceptibility to infection and virus gene expression increase during maturation of monocytes to macrophages. *J. Virol.*, **58**, 67–74.

Gerhard, D. S. *et al.* (1981). Localization of a unique gene by direct hybridization *in situ*. *Proc. Natl Acad. Sci. USA*, **78**, 3755–9.

Geukens, M. and May, E. (1974). Ultrastructural localization of SV40 viral DNA in cells, during lytic infection, by *in situ* molecular hybridization. *Exp. Cell. Res.*, **87**, 175–85.

Gomes, S. A., Nascimento, J. P., Siquera, M. M., Krawczuk, M. M., Pereira, H. G., and Russell, W. C. (1985). *In situ* hybridization with biotinylated DNA probes: a rapid diagnostic test for adenovirus upper respiratory infections. *J. Virol. Methods.*, **12**, 105–10.

Gowans, E. J., Burrell, C. J., Jilbert, A. R., and Marmion, B. P. (1981). Detection of hepatitis B virus DNA sequences in infected hepatocytes by *in situ* cyto-hybridisation. *J. Med. Virol.*, **8**, 67–78.

Grussendorf-Conen, E. I., Ikenberg, H., and Gissman, L., (1985). Demonstration of HPV-16 genomes in the nuclei of cervix carcinoma cells. *Dermatologica*, **170**, 199–201.

Gupta, J. W., Gupta, P. K., Rosenshein, N., and Shah, K. (1987). Detection of human papillomavirus in cervical smears. *Acta. Cytol.*, **8**, 387–95.

Gupta, J. *et al.* (1985). Specific identification of human papillomavirus type in cervical smears and paraffin sections by *in situ* hybridization with radioactive probes: a preliminary communication. *Int. J. Gyn. Pathol.*, **4**, 211–18.

Haase, A. T., Stowring, L., Narayan, O., Griffin, D., and Price, D. (1977). Slow persistent infection caused by visna virus: role of host restriction. *Science*, **195**, 175–7.

Haase, A. T., Ventura, P., Gibbs, C. J., and Tourtellotte, W. W. (1981*a*). Measles virus nucleotide sequences: detection by hybridization *in situ*. *Science*, **212**, 672–5.

Haase, A. T. *et al.* (1981*b*). Measles virus genome in infections of the central nervous system. *J. Inf. Dis.*, **144**, 154–60.

Haase, A. T. *et al.* (1985*a*). Combined macroscopic and microscopic detection of viral genes in tissues. *Virology*, **140**, 201–6.

Haase, A. T. *et al.* (1985*b*). Detection of two viral genomes in single cells by double-label hybridization *in situ* and color microradioautography. *Science*, **227**, 189–92.

Haase, A. T. *et al.* (1985*c*). Natural history of restricted synthesis and expression of measles virus genes in subacute sclerosing panencephalitis. *Proc. Natl. Acad. Sci. USA*, **82**, 3020–4.

Harper, M. E., Ullrich, A., and Sauders, G. F. (1981). Localization of the human insulin gene to the distal end of the short arm of chromosome 11. *Proc. Natl Acad. Sci. USA*, **78**, 4458–60.

Harper, M. E., Marselle, L. M., Gallo, R. C., and Wong-Staal, F. (1986). Detection of lymphocytes expressing human T-lymphotropic virus type III in lymph nodes and peripheral blood from infected individuals by *in situ* hybridization. *Proc. Natl Acad. Sci. USA*, **83**, 772–6.

Harris, A., Young, B., and Griffin, B. E. (1985). Random association of Epstein-Barr virus genomes with host cell metaphase chromosomes in Burkitt's lymphoma-derived cell lines. *J. Virol.*, **56**, 328–32.

Hauser, S. L. *et al.* (1986). Analysis of human T-lymphotropic virus sequences in multiple sclerosis tissue. *Nature*, **322**, 176–7.

Henderson, A. S., Yu, M. T., and Silverstein, S. (1981). Chromosomal DNA homologous to Herpes simplex virus 1 in a mouse L-cell line. *Cytogenet. Cell. Genet.*, **29**, 107–15.

Henderson, A., Ripley, S., Heller, M., and Kieff, E. (1983). Chromsome site for Epstein-Barr virus DNA in a Burkitt tumor cell line and in lymphocytes growth-transformed *in vitro*. *Proc. Natl Acad. Sci. USA*, **80**, 1987–91.

Hirai, J., Maotani, K., Ikuta, K., Yasue, H., Ishibashi, M., and Kato, S. (1986). Chromosomal sites for Marek's disease virus DNA in two chicken lympho-blastoid cell lines MDCC-MSB1 and MDCC-RP1. *Virology*, **152**, 256–61.

Horvath, J., Palkonyay, L., and Weber, J. (1986). Group C adenovirus DNA sequences in human lymphoid cells. *J. Virol.*, **59**, 189–92.

Houff, S. A. *et al.* (1988). Involvement of JC virus-infected mononuclear cells from the bone marrow and spleen in the pathogenesis of progressive multifocal leukoencephalopathy. *N. Engl. J. Med.*, **318**, 301–5.

Hyman, R. W., Ecker, J. R., and Tenser, R. B. (1983). Varicella-zoster virus RNA in human trigeminal ganglia. *Lancet*, **2**, 814–16.

Ishibashi, M., Yosida, T. H., and Yasue, H. (1987). Preferential clustering of viral DNA sequences at or near the site of chromosomal rearrangement in fowl adenovirus type 1 DNA-transformed cell lines. *J. Virol.*, **61**, 151–8.

Janssen, H. P., van Loon, A. M., Meddens, M. J., Herbrink, P., Lindeman, J., and Quint, W. G. (1987). Comparison of *in situ* DNA hybridization and immuno-logical staining with conventional virus isolation for the detection of human cytomegalovirus infection in cell cultures. *J. Virol. Methods*, **17**, 311–18.

Jarret, W. F. H. (1985). The natural history of bovine papillomavirus infections. *Adv. Viral Oncol.* **5**, 83–101.

Jilbert, A. R. *et al.* (1987). Duck hepatitis B virus DNA in liver, spleen and pancreas: analysis by *in situ* and Southern blot hybridization. *Virology*, **158**, 330–8.

John, H. L., Birnstiel, M. I., and Jones, K. W. (1969). RNA–DNA hybrids at the cyto-logical level. *Nature*, **223**, 582–4.

Jones, J. F. *et al.* (1988). T-cell lymphomas containing Epstein–Barr viral DNA in patients with chronic Epstein–Barr virus infections. *N. Engl. J. Med.*, **318**, 733–41.

Jones, K. W., Kinross, L., and Maitland, N. (1979). Normal human tissues contain RNA and antigens related to infectious adenovirus type 2. *Nature*, **277**, 274–9.

Kandolf, R., Ameis, D., Kirschner, P., Canu, A., and Hofschneider, P. H., (1987). *In situ* detection of enteroviral genomes in myocardial cells by nucleic acid hybridization: an approach to the diagnosis of viral heart disease. *Proc. Natl. Acad Sci. USA*, **84**, 6272–6.

Kaufman, S. L., Gallo, R. C., and Miller, N. R. (1979). Detection of virus-specific RNA in simian sarcoma-leukemia virus-infected cells by *in situ* hybridization to viral complementary DNA. *J. Virol.*, **30**, 637–41.

Keh, W. C., and Gerber, M. A. (1988). *In situ* hybridization for cytomegalovirus DNA in AIDS patients. *Am. J. Pathol.*, **131**, 490–6.

Koprowski, H. *et al.* (1985). Multiple sclerosis and human T-cell lymphotropic retroviruses. *Nature*, **318**, 154–60.

Korba, B. E., Gowans, E. J., Wells, F. V., Tennant, B. C., Clark, R., and Gerin, J. L. (1988). Systemic distribution of woodchuck hepatitis virus in the tissues of experimentally infected woodchucks. *Virology*, **165**, 172–81.

Kurtzman, G. J., Ozawa, K., Cohen, B., Hanson, G., Oseas, R., and Young, N. S. (1987). Chronic bone marrow failure due to persistent B19 parvovirus infection. *N. Engl. J. Med.*, **317**, 287–94.

Langenberg, A. *et al.* (1988). Detection of herpes simplex virus DNA from genital lesions by *in situ* hybridization. *J. Clin. Microbiol.*, **26**, 933–7.

Law, M.-F., Lowy, D. R., Dvoretzky, I., and Howley, P. (1981). Mouse cells transformed by bovine papillomavirus contain only extrachromosomal viral DNA sequences. *Proc. Natl Acad. Sci. USA*, **78**, 2727–31.

Lawrence, J. B., Villnave, C. A., and Singer, R. H. (1988). Sensitive, high-resolution chromatin and chromosome mapping *in situ*: presence and orientation of two closely integrated copies of EBV in a lymphoma line. *Cell*, **52**, 51–6.

Loni, M. C. and Green, M. (1974). Detection and localization of virus-specific DNA by *in situ* hybridization of cells during infection and rapid transformation by the murine sarcoma-leukemia virus. *Proc. Natl Acad. Sci. USA.*, **71**, 3418–22.

Loning, T., Milde, K., and Foss, H.-D. (1986). *In situ* hybridization for the detection of cytomegalovirus (CMV) infection. *Virchows Arch.*, **409**, 777–90.

Maitland, N. J., Kinross, J. H., Busuttil, A., Holgate, S. M., Smart, G. E., and Jones, K. W., (1981). The detection of DNA tumour virus-specific RNA sequences in abnormal human cervical biopsies by *in situ* hybridization. *J. Gen. Virol.*, **55**, 123–37.

Mason, W. S. *et al.* (1982). Asymmetric replication of duck hepatitis B virus DNA in liver cells: free minus-strand DNA. *Proc. Natl Acad. Sci. USA*, **79**, 3997–4001.

McDougall, J. K., Dunn, A. R., and Jones, K. W. (1972). *In situ* hybridization of adenovirus RNA and DNA. *Nature*, **236**, 346–8.

McDougall, J. K., Galloway, D. A., and Fenoglio, C. M. (1980). Cervical carcinoma: detection of herpes simplex virus RNA in cells undergoing neoplastic change. *Int. J. Cancer*, **25**, 1–8.

Mercer, J. A., Clayton, A. W., and Spector, D. H. (1988). Pathogenesis of murine cytomegalovirus infection: identification of infected cells in the spleen during acute and latent infections. *J. Virol.*, **62**, 987–97.

Myerson, D., Hackman, R. C., Nelson, J. A., Ward, D. C., and McDougall, J. K. (1984*a*). Widespread presence of histologically occult cytomegalovirus. *Hum. Pathol.*, **15**, 430–9.

Myerson, D., Hackman, R. C., and Meyers, J. D. (1984*b*). Diagnosis of cyto-megalovirus pneumonia by *in-situ* hybridization. *J. Infect. Dis.*, **150**, 272–7.

Nakane, P. K., Moriuchi, T., Koji, T., Tanno, M., and Abe, K. (1987). *In situ* localization of mRNA using thymidine–thymidine dimerized cDNA. *Acta Histochem. Cytochem.*, **20**, 229–43.

Neumann, R., Gensersch, E., and Eggers, H. J. (1987). Detection of adenovirus nucleic acid sequences in human tonsils in the absence of infectious virus. *Virus Res.*, **7**, 93–7.

Orth, G., Jeanteur, P., and Croissant, O. (1970). Evidence for and localization of vegetative viral DNA replication by autographic detection of RNA-DNA hybrids in sections of tumours induced by Shope papilloma virus. *Proc. Natl Acad. Sci. USA.*, **68**, 1876–81.

Ostrow, R. S., Manias, D. A., Clark, B. A., Okagaki, T., Twiggs, L. B., and Faras, A. J. (1987). Detection of human papillomavirus DNA in invasive carcinomas of the cervix by *in situ* hybridization. *Cancer Res.*, **47**, 649–53.

Pardue, M. L., and Gall, J. G. (1969). Chromosomal localization of mouse satellite DNA. *Science*, **168**, 1356–8.

Peluso, R., Haase, A., Stowring, L., Edwards, M., and Ventura, P. (1985). A Trojan horse mechanism for the spread of visna virus in monocytes. *Virology*, **147**, 231–6.

Popescu, N. C., DiPaolo, J. A., and Amsbaugh, S. C. (1987*a*). Integration sites of human papilomavirus 18 DNA sequences on HeLa cell chromosomes. *Cytogenet. Cell Genet.*, **44**, 58–62.

Popescu, N. C., Amsbaugh, S. C., and DiPaolo, J. A. (1987*b*). Human papillomavirus type 18 DNA is integrated at a single chromosome site in cervical carcinoma cell line SW756. *J. Virol.*, **51**, 1682–5.

Porter, H. J., Khong, T. Y., Evans, M. F., Chan, V. T.-W., and Fleming, K. A. (1988). Parvovirus as a cause of hydrops fetalis: detection by *in situ* hybridisation. *J. Clin. Pathol.*, **41**, 381–3.

Raap, A. K., Geelen, J. L., van der Meer, J. W. M., van de Rijke, F. M., van den Boogart, P., and van der Ploeg, M. (1988). Non radioactive *in situ* hybridization for the detection of cytomegalovirus infections. *Histochemistry*, **88**, 367–73.

Rabin, M., Uhlenbeck, O. C., Steffensen, D. M., and Mangel, W. F. (1984). Chromosomal sites of integration of simian virus 40 DNA sequences mapped by *in situ* hybridization in two transformed hybrid cell lines. *J. Virol.*, **49**, 445–51.

Rappolee, D. A., Mark, D., Banda, M. J., and Werb, Z. (1988). Wound macrophages express TGF-alpha and other growth factors *in vivo*: analysis by mRNA phenotyping. *Science*, **241**, 708–12.

Rein, A., Rice, N., Simek, S., Cohen, M., and Mural, R. (1982). *In situ* hybridization: general infectivity assay for retroviruses. *J. Virol.*, **43**, 1055–60.

Rock, D. L. *et al.* (1987). Detection of latently-related viral RNAs in trigeminal ganglia of rabbits latently infected with herpes simplex virus type 1. *J. Virol.*, **61**, 3820–6.

Ross, N. L. J., deLorbe, W., Varmus, H. E., Bishop, J. M., Brahic, M., and Haase, A. (1981). Persistence and expression of Marek's disease virus DNA in tumour cells and peripheral nerves studied by *in situ* hybridization. *J. Gen. Virol.*, **57**, 285–96.

Schneider, A., Oltersdorf, T., Schneider, V., and Gissman, L. (1987). Distribution pattern of human papilloma virus 16 genome in cervical neoplasia by molecular *in situ* hybridization of tissue sections. *Int. J. Cancer*, **39**, 717–21.

Schrier, R. D., Nelson, J. A., and Oldstone, M. B. A. (1985). Detection of human cytomegalovirus in peripheral blood lymphocytes in a natural infection. *Science*, **230**, 1048–51.

Sharpsak, P. *et al.* (1986). Search for virus nucleic acid sequences in postmortem human brain tissue using *in situ* hybridization technology with cloned probes: some solutions and results on progressive multifocal leukoencephalopathy and subacute sclerosing panencephalitis tissue. *J. Neurosci. Res.*, **16**, 281–301.

Shaw, G. M. *et al.* (1985). HTLV-III infection of brains of children and adults with AIDS encephalopathy. *Science*, **227**, 177–82.

Shiraishi, Y. *et al.* (1985). Chromosomal localization of the Epstein-Barr virus (EBV) genome in Bloom's syndrome B-lymphoblastoid cell lines transformed with EBV. *Chromosoma (Berlin)*, **93**, 157–64.

Simon, I., Lohler, J., and Jaenisch, R. (1982). Virus-specific transcription and translation in organs of BALB/Mo mice: comparative study using quantitative hybridization, *in situ* hybridization and immunocytochemistry. *Virology*, **129**, 106–21.

Simon, D. *et al.* (1985). Chromosomal site of hepatitis B virus (HBV) integration in a human hepatocellular carcinoma-derived cell line. *Cytogenet. Cell. Genet.*, **39**, 116–20.

Sixbey, J. W., Nedrud, J. G., Raab-Traub, N., Hanes, R. A., and Pagano, J. S. (1984). Epstein-Barr virus replication in oropharyngeal epithelial cells. *N. Engl. J. Med.*, **310**, 1225–30.

Skare, J. and Strominger, J. (1980). Cloning and mapping of BamHI endonuclease fragments of DNA of the transforming B95-8 strain of EBV. *Proc. Natl Acad. Sci. USA*, **77**, 3860–4.

Steiner, I., Spivack, J. G., Linette, R. P., Brown, S. M., McLearn, A. R., Subak-Sharpe, J. H., and Fraser, N. W. (1989). Herpes simplex virus type I. Latency associated transcripts are evidently not essential for latent infection. *EMBO J.*, **8**, 505–11.

Stevens, J. G., Wagner, E. K., DeviRao, G. B., Cook, M. L., and Feldman, L. T. (1987). RNA complementary to a herpesvirus alpha gene mRNA is prominent in latently infected neurons. *Science*, **235**, 1056–9.

Stevens, J. G., Haarr, L., Porter, D. D., Cook, M. L., and Wagner, E. K. (1988). Prominence of the herpes simplex virus latency-associated transcript in trigeminal ganglia from seropositive humans. *J. Infect. Dis.*, **158**, 117–23.

Stoler, M. H., and Broker, T. R. (1987). *In situ* hybridization detection of human papillomavirus DNAs and messenger RNAs in genital condylomas and a cervical carcinoma. *Human Pathol.*, **17**, 1250–8.

Stowring, L. *et al.* (1985). Detection of visna virus antigens and RNA in glial cells in foci of demyelination. *Virology*, **141**, 311–18.

Stroop, W. G., Rock, D. L., and Fraser, N. W. (1984). Localization of herpes simplex virus in the trigeminal and olfactory systems of the mouse central nervous system during acute and latent infections by *in situ* hybridization. *Lab Invest.*, **51**, 27–3.

Tecott, L. H., Barchas, J. D., and Eberwine, J. G. (1988). *In situ* transcription: specific synthesis of complementary DNA in fixed tissue sections. *Science*, **240**, 1661–4.

Teo, C. G. and Griffin, B. E. (1987). Epstein–Barr virus genomes in lymphoid cells: activation in mitosis and chromosomal location. *Proc. Natl Acad. Sci. USA*, **84**, 8473–7.

Teo, C. G., Wong, S. Y., and Best P. V. (1989). JC virus genomes in progressive multifocal leukoencephalopathy: detection using a sensitive non-radioisotopic *in situ* hybridization method. *J. Pathol.*, **157**, 135–40.

Trescol-Biemont, M.-C., Biemont, C., and Daille, J. (1987). Localization polymorphism of EBV DNA genomes in the chromosomes of Burkitt lymphoma cell lines. *Chromosoma (Berlin)*, **95**, 144–7.

Unger, E. R., Budgeon, L. R., Myerson, D., and Brigati, D. J. (1986). Viral diagnosis by *in situ* hybridization. *Am. J. Surg. Pathol.*, **10**, 1–8.

Vafai, A., Murray, R. S., Wellish, M., Devlin, M., and Gilden, D. H. (1988). Expression of varicella-zoster virus and herpes simplex virus in normal human trigeminal ganglia. *Proc. Natl Acad. Sci. USA*, **85**, 2362–6.

Venables, P. J. W., Teo, C. G., Baboonian, C., Griffin, B. E., Hughes, R. A., and Maini, R. N. (1989). Persistence of Epstein-Barr virus in salivary gland biopsies from healthy individuals and patients with Sjogren's syndrome. *Clin. Exp. Immunol.*, **75**, 359–64.

Vogel, S., Rosahl, T., and Doerfler, W. (1986). Chromosomal localization of integrated adenovirus DNA in productively infected and in transformed mammalian cells. *Virology*, **152**, 159–70.

Wang, L., Kemp, M. C., Roy, P., and Collison, E. W. (1988). Tissue tropism and target cells of bluetongue virus in the chicken embryo. *J. Virol.*, **62**, 887–93.

Wechsler, S. L., Nesburn, A. B., Watson, R., Slanina, S. M., and Ghiasi, H. (1988). Fine mapping of the latency-related gene of herpes simplex virus type 1: alternative splicing produces distinctly latency-related RNAs containing open reading frames. *J. Virol.*, **62**, 4051–8.

Wolf, H., zur Hausen, H., and Becker, V. (1973). EB viral genomes in epithelial nasopharyngeal carcinoma cells. *Nature New Biol.*, **244**, 245–7.

Wolf, H., Haus, M., and Wilmes, E. (1984). Persistence of Epstein–Barr virus in the parotid gland. *J. Virol.*, **51**, 795–8.

zur Hausen, H. and Schulte-Holthausen, H. (1972). Detection of Epstein-Barr viral genomes in human tumour cells by nucleic acid hybridization. In *Oncogenesis and herpesvirus*, (ed. P. M. Biggs and L. N. Payne), pp. 321–5. IARC, Lyon.

zur Hausen, H., Meinhof, W., Scheiber, W., and Bornkamm, G. W. (1974). Attempts to detect virus-specific DNA in human tumours. 1. Nucleic acid hybridizations with complementary RNA of human wart virus. *Int. J. Cancer.*, **13**, 650–6.

10

Chromosomal assignment of genes

B. BHATT and J. O'D. McGEE

Introduction

The history of *in situ* hybridization dates back to 1969 when Gall and Pardue demonstrated the detection of amplified ribosomal RNA genes in the nuclei of *Xenopus* oocytes. In recent years, it has become a powerful tool for the detection of nucleic acids in individual cells, tissue sections, and chromosomes. *In situ* hybridization involves hybridization of labelled nucleic acid probe with target nucleic acids in a cell or tissue section. If the probe is labelled with a radioisotope, the product is detected by auto-radiography (Pardue 1985; Buckle and Craig 1986). On the other hand, non-radioactive labels such as biotin (Bhatt *et al.* 1988; Garson *et al.* 1987) or 2-acetylaminofluorine (Landegent *et al.* 1985) are detected by immuno-cytochemical methods. This technology has been exploited in detecting viral DNA sequences in infected tissue (Brahic *et al.* 1984; Brahic and Hasse 1978; Burns *et al.* 1987, 1988; Herrington *et al.* 1989*a,b,c,d*; see Chapter 12); mRNA expression (Shivers *et al.* 1986; Terenghi *et al.* 1987; Hamid *et al.* 1987) and in the regulation of gene expression (Tank *et al.* 1985; Steel *et al.* 1988); antenatal sex determination (Burns *et al.* 1985; Julien *et al.* 1986); cell ontogeny in therapeutic human chimeras (Lyttleton *et al.* 1988; Reittie *et al.* 1988, Athanason *et al.* 1990); and human gene mapping (Harper and Saunders 1981; Bhatt *et al.* 1988).

Genes and DNA sequences of interest can be mapped to their respective chromosomes by several methods. The two most popular methods are somatic cell hybridization (Antonarakis *et al.* 1983) and *in situ* hybridiz-ation (Zabel *et al.* 1985). In the former, labelled DNA sequences of interest are hybridized to DNA from somatic cell hybrids carrying different human chromosomes. By contrast, *in situ* hybridization involves hybridization of labelled probe to metaphase spreads, and so permits direct visualization of genes on human chromosomes. This not only provides a physical basis for the human gene map, but also provides an important reference point for cloning and molecular dissection of genes of interest such as those involved in genetic diseases and cancer.

Until 3 or 4 years ago, single-copy genes were detected by probes labelled with ^{3}H or ^{32}P followed by autoradiography. In the last 3–4 years, several laboratories have been attempting to find non-isotopic alternatives that can match the sensitivity and resolution afforded by tritium. Some of the alternative labels employed include biotin and 2-acetylaminofluorene (AAF). The former may be incorporated in the probe by conventional nick translation or by chemical modification (Forster *et al.* 1985), and the latter by chemical modification (Landegent *et al.* 1985). Whatever the label, non-radioactive reporter molecules are invariably detected by histochemical techniques. In the authors' laboratory, biotin has been successfully exploited in mapping genes with probes as small as 0.8–1 kb. Although the methodology described here applies mainly to biotin-labelled probes it can, in principle, with minor modifications, be applied to other reporter molecules such as photobiotin, digoxigenin (see Chapter 12) and bromo-deoxyuridine (BUdR). The latter two reporters have still to be fully evaluated for their use as labels for chromosomal gene assignment. Pre-liminary experiments in this laboratory indicate that digoxigenin can be used for this purpose.

The facility of non-isotopic *in situ* hybridization (NISH) for chromo-somal gene assignment will be illustrated by six examples; the assignment of NRAS, parathyroid hormone (PTH), β-globin, HRAS1, insulin, and α-globin. Before discussing these assignments, however, there are several technical points that need to be stressed. The Appendix deals with the entire methodology used in this laboratory.

Methodological considerations

The starting point for chromosomal gene assignments by NISH are good metaphase spread that can be banded after hybridization. This allows the gene of interest to be visualized directly on banded chromosomes. In outline, the technique used in this laboratory is as follows.

Metaphase spreads are prepared from peripheral blood lymphocytes that have been pulsed with BUdR, which is necessary for replication banding of chromosomes. This approach allows single-copy genes to be visualized directly on banded chromosomes by regular light microscopy in a single step. The metaphases are then hybridized with biotinylated probes and after stringency washes, the biotinyl residues are detected by an antibody procedure. The immunohistochemical signal is further amplified by silver precipitation. The chromosomes are replication-banded. The gene signal is visualized directly on these banded chromosomes as a black silver dot. The latter forms an excellent contrast on the red/blue banded chromosomes (see Figures). Each of the parts of the procedure, and possible pitfalls, are discussed below before the results of applying the methodology are presented.

Preparation of metaphase spreads and chromosome banding

Whole blood cultures provide an abundant source of material for routine cytogenetics and gene mapping. Fixation of cells with methanol/acetic acid not only fixes the cells but also ensures proper spreading of chromosomes on slides. Replication banding enables the NISH signal (silver dots) to be visualized directly on banded chromosomes (Bhatt *et al.* 1988).

Alternative banding methods such as trypsin-Giemsa (Seabright 1972) or acetic saline Giemsa (Sumner *et al.* 1971) may be employed for banding. However, it should be borne in mind that the latter banding procedures destroy the signal. Using these banding procedures, therefore, it is necessary to photograph the spreads with signal on unbanded chromosomes (Garson *et al.* 1987). The same spreads must then be rephotographed after chromosome banding, and the two sets of photographs compared to localize the signal on chromosome bands.

Replication bands are not synonymous with G-bands, although after synchronization with BUdR most metaphase spreads will show a banding pattern more or less identical to G-bands.

Labelling, denaturation, and hybridization

The labelling protocol is identical to that of radioactive labelling except for the substitution of the radioisotope with biotin-11-dUTP or biotin-7-dATP. A series of probe concentrations should be tried. Higher probe concentrations (up to 200 ng per 6 cm^2 slide) will ensure that the probe can complex effectively with the gene before rapid reannealing of chromosomal DNA which occurs after denaturation (Singer *et al.* 1987).

The stability of DNA–DNA or RNA–DNA hybrids can be assessed from the melting temperature (T_m). T_m is the temperature at which 50% of the nucleic acid duplexes became dissociated (see Chapter 1). T_m is calculated from the following equation.

$$T_m = 81.5 + 16.6 \ (\log M) + 0.41 \ (\%G + C) - 650/L - 0.72F - 1.4 \ (\% \text{ mismatch})$$

where M = ionic strength (mol/litre), L = probe length, F = % formamide.

The DNA–DNA reassociation temperature (T_r) occurs 25 °C below T_m. Lower salt concentration and higher temperature favour accurate base pairing. However, the presence of formamide lowers the temperature required for accurate base pairing and so helps preserve the morphology of chromosomes. Dextran sulphate has been shown to increase the rate of reassociation of DNA (Wahl *et al.* 1979). Human DNA or salmon sperm DNA are included in the hybridization mix to reduce non-specific binding of labelled probe.

The optimal temperatures, salt concentration etc., for denaturation and hybridization at high stringency for unique sequences are given in the Appendix. Depending on the object of the experiment, these conditions may have to be modified (see below).

Stringency washes

With RNA–DNA hybrids, non-specifically bound RNA can be removed by treatment with RNaseA. In case of DNA–DNA hybrids, non-specifically bound DNA probe cannot be removed by enzymatic digestion. It should, therefore, be removed by washes in salt solutions of varying ionic strengths, or by incubation in low salt ($\pm$ formamide) at high temperatures. Incubation at high temperatures for prolonged periods destroys the architecture of the chromosomes and should be avoided. Incubation in 50% formamide in $2 \times SSC$; or $0.1 \times SSC$ for even higher stringency at a temperature 5–7 °C higher than the hybridization temperature will 'melt' non-specifically bound probe DNA. This, followed by washes in $2 \times SSC$ and $0.1 \times SSC$, will ensure the removal of most of the non-specifically bound DNA and ensure a higher signal-to-noise ratio. Signal-to-noise ratios of 10:1 and higher are routinely achieved.

Immunocytochemical detection of bound probe

Incubation of slides in buffered saline containing bovine serum albumin blocks non-specific antibody binding sites. Tris-buffered saline may be used instead of phosphate-buffered saline. Singer *et al.* (1987) suggested that where avidin is used to detect biotinylated probe, $4 \times SSC$ gives less background noise than phosphate buffered saline. Avidin conjugated to gold particles of varying sizes may also be used. Although the affinity of rabbit antibiotin for biotin is less than avidin, the increase in the number of steps amplifies the signal. A monoclonal antibiotin is now available (Dako, UK). The second antibody is a goat anti-rabbit IgG (GARIgG) conjugated to gold particles of 5 nm, 10 nm, or 30 nm. If further amplification is desired, GARIgG not conjugated with gold may be used. Two to three cycles of primary antibody followed by second antibody amplifies the signal substantially. GARIgG conjugated with gold should be employed in the last cycle. Because the second antibody is raised in goat, non-specific antibody-binding sites may be blocked by incubation in PBS containing 5 per cent normal goat serum prior to applying the second antibody.

Before applying the silver solution, it should be made absolutely certain that no phosphate ions remain on the spreads, because the phosphates will precipitate silver. The silver solution comes in the form of two components called initiator and enhancer. These should be mixed immediately before use. The manufacturer's recommendations should be followed.

Gene assignment

NRAS

Figure 10.1(a) shows a metaphase spread with a silver dot on band 1p1.34, where the *N-ras* gene is located. Unlike some repeat probes such as pHY2.1 which labels the Y chromosome (Fig. 10.2), some background will inevitably occur when more complex unique sequence genes are probed (Fig. 10.1(a)). This is because the probe itself hybridizes non-specifically to other chromosomal regions and the immunocytochemical detection system also causes some background. It is therefore, essential to count silver dots on about 50 metaphase spreads and plot them on a G-band karyogram, as shown in Fig. 10.1(b). The chromosomal band that shows a statistically significant clustering of silver dots is judged as the band that carries the gene under investigation (Fig. 10.1(c)).

Parathyroid hormone, HRAS1, β-globin, insulin

More recently, we have attempted to work out the spatial order of four genes on the short arm of chromosome 11, viz. on band p15. The four genes studied were β-globin, HRAS1, PTH, and insulin. As shown in Fig. 10.3, it is possible to resolve, discretely, the PTH locus (junction of p14/p15) and the insulin locus which is the most distal of these genes at the tip of p15. Although it is likely that β-globin lies proximal to insulin and distal to HRAS1 on p15, the resolution of the present technology is not sufficiently discriminative to be certain of this. From these experiments, it is concluded that the gene order on p15 is PTH, HRAS1, β-globin, and insulin. The inability to achieve firmer resolution is dependent on the degree of extension of the chromosome of interest and, more importantly, the size of the silver grain found at the gene locus (for detailed discussion see Bhatt *et al.* 1988).

α-Globin and gene families

NISH assignment of genes not only identifies the gene of interest but also partially homologous genes. For example, Bhatt *et al.* (1988) showed that under low stringency the NRAS probe also reacted with the HRAS1 and KRAS locus on chromosomes 11 and 12, respectively. Of equal interest was the finding that α-globin in NISH assignment also identified a partially homologous sequence on chromosome 11q (Nicholls *et al.* 1987). It has since been confirmed that there is a partially homologous sequence to α-globin on this chromosome by independent forms of analysis. These observations raise the possibility that chromosomal assignment of gene families (and the discovery, therefore, of hitherto unknown sequences) can be identified by NISH under conditions of low stringency.

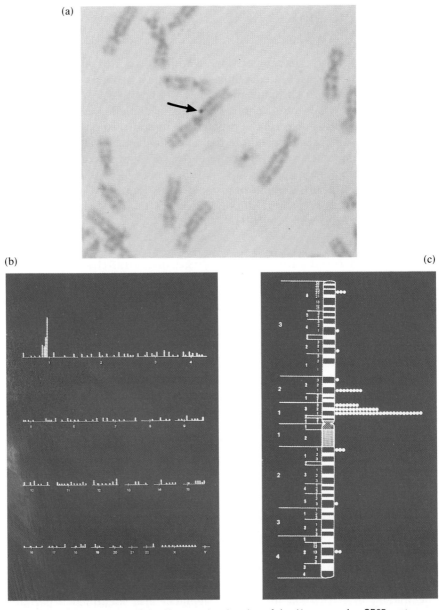

Fig. 10.1. A 1.5 kb fragment of the 3′ untranslated region of the *N-ras* gene in pSP65 vector was labelled with biotin and hybridized to human mitotic chromosomes. (a) A silver dot (arrow) is present on band 1p13. (b) Histogram showing the distribution of 252 silver dots on 60 meta-phase spreads. The significant cluster of dots on chromosome 1p constitutes the signal. The presence of dots over the rest of the chromosomes in metaphase spreads forms the background noise. (c) The distribution of silver dots on chromosome 1. (Reproduced with permission from Bhatt *et al.* (1988). *Nucl. Acids Res.* **16**, 3951–61. IRL Press Ltd, Oxford, UK.)

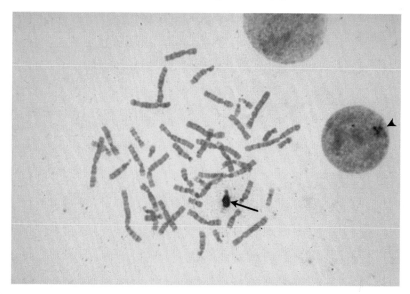

Fig. 10.2. pHY2.1 in pBR328 vector, a Y-chromosome-specific repeat was labelled with biotin and hybridized to male metaphase spreads. The distal half of the long arm of chromosome Y (Yq12), which is entirely composed of heterochromatic DNA, labels with this probe (arrow); 95–99 per cent of the spreads can be labelled, with little or no background noise. A signal is also present in an interphase nucleus (arrowhead).

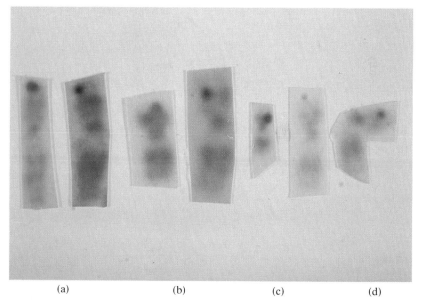

| (a) | (b) | (c) | (d) |

Fig. 10.3. *In situ* hybridization of (a) insulin, (b) β-globin, (c) PTH, and (d) H-RAS1 probes to metaphase spreads. The insulin insert (biotinylated) was 0.8 kb.

In situ *hybridization*

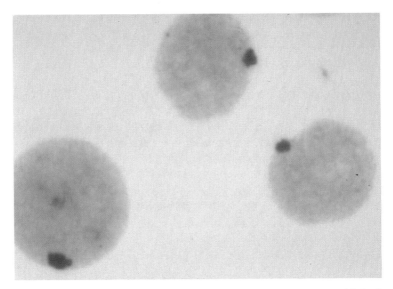

Fig. 10.4. Male lymphocytes probed with biotin-labelled pHY2.1 (see Fig. 10.2). All the lymphocytes contain a single black signal representing the long arm of the Y chromosome.

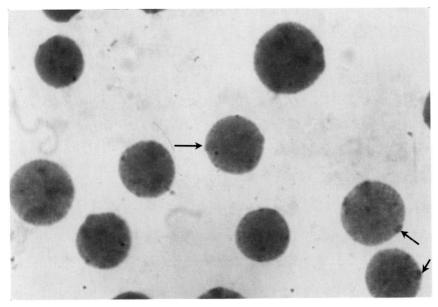

Fig. 10.5. Male human lymphocytes (metaphase spreads) hybridized with the probe for the hypoxanthine guanine phosphoribosyl transferase (HPRT) gene. The nuclei show single dots (arrows). Cells were Giemsa counterstained.

Interphase cytogenetics

Repeat sequences, specific for any given chromosomes, can be easily identified by NISH in interphase cells (Fig. 10.4). When probes become available for each individual chromosome it will be possible to analyse the entire chromosome complement of interphase nuclei (see Herrington *et al.* 1989*a,b*; 1990*a,b* for a full discussion; see also Chapter 11).

One of the goals of this laboratory is to refine interphase cytogenetic analysis by NISH to a level at which unique sequences can be identified in interphase nuclei. Preliminary data (Bhatt and McGee, unpublished) indicate that unique sequences can be visualized in interphase nuclei of lymphocytes (Figs. 10.5 and 10.6). However, at the present time, it is not possible to identify unique sequences in every cell (in an entire population)

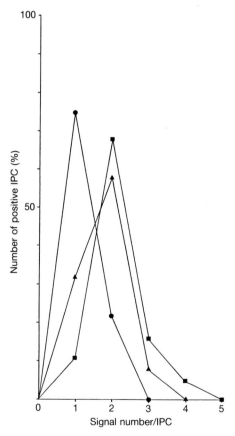

Fig. 10.6. Human male lymphocytes hybridized with probes for HPRT, N-RAS and α-1-antitrypsin (AAT) genes. 100 cells were counted for each probe. There is on average 1 signal per cell (IPC) for HPRT (this gene is on the X chromosome) and 2 signals for NRAS and AAT, which are on chromosomes 1 and 14, respectively. (a) ●, HPRT; (b) ■; N-RAS; (c) ▲, AAT.

without background (Fig. 10.6). These difficulties should be overcome in the near future.

Interphase cytogenetic analysis will have wide application in antenatal diagnosis (with oligonucleotide allele-specific probes) and in the analysis of gene deletions/anomalies in human cancer biopsies.

References

Antonarakis, S. E., Phillips, J. A., III, Mallonee, R. L., Kazazia, H. H., Jr, Fearon, E. R., Waber, P. G., Kronenberg, H. M., Ullrich, A., and Meyers, D. A. (1983). β-globin locus is linked to the parathyroid hormone (PTH) locus and lies between the insulin and PTH loci in man. *Proc. Natl Acad. Sci. USA*, **80**, 6615–19.

Athanason, N. A., Quinn, J., Brenner, M. K., Prentice, G. H., Graham, A., Taylor, S., Flannery, D., and McGee, J. O'D. (1990). Origin of marrow stromal cells and haemopoietic chimaerism following bone marrow transplantation determined by *in situ* hybridisation. *Br. J. Cancer.* (In press.)

Bhatt, B., Burns, J., Flannery, D., and McGee, J. O'D. (1988). Direct visualisation of single copy genes on banded metaphase chromosomes by nonisotopic *in situ* hybridisation. *Nucl. Acids Res.*, **16**, 3951–61.

Brahic, M. and Hasse, A. T. (1978). Detection of viral sequences of low reiteration frequency by *in situ* hybridisation. *Proc. Natl Acad. Sci. USA*, **75**, 6125–9.

Brahic, M., Hasse, A. T., and Cash, E. (1984). Simultaneous *in situ* detection of viral RNA and antigens. *Proc. Natl Acad. Sci. USA*, **81**, 5445–8.

Buckle, V. and Craig, I. W. (1986). *In situ* hybridisation. In *Human genetic diseases. A practical approach*, (ed. K. E. Davies). IRL Press, Oxford.

Burns, J., Chan, V. T.-W., Jonasson, J. A., Fleming, K. A., Taylor, S., and McGee, J. O'D. (1985). Sensitive system for visualising biotinylated DNA probes hybridised *in situ*: rapid sex determination of intact cells. *J. Clin. Pathol.*, **38**, 1085–92.

Burns, J., Graham, A. K., Frank, C., Fleming, K. A., Evans, M. P., and McGee, J. O'D. (1987). Detection of low copy human papilloma virus DNA and mRNA in routine paraffin sections of cervix by nonisotopic *in situ* hybridisation. *J. Clin. Pathol.*, **40**, 858–64.

Forster, A. C., McInnes, J. L., Skingle, D. C., and Symons, R. H. (1985). Non-radioactive hybridisation probes prepared by the chemical labelling of DNA and RNA with a novel reagent, photobiotin. *Nucl. Acids Res.*, **13**, 745–61.

Gall, J. G. and Pardue, M. (1969). Formation and detection of RNA–DNA hybrid molecules in cytological preparations. *Proc. Natl Acad. Sci. USA*, **63**, 378–83.

Garson, J. A., van den Berghe, J. A., and Kemshead, J. T. (1987). Novel nonisotopic *in situ* hybridisation technique detects small (1 kb) unique sequences in routinely G-banded human chromosomes: fine mapping of N-myc and b-NGF genes. *Nucl. Acids Res.*, **15**, 4761–70.

Hamid, Q., Wharton, J., Terenghi, G., Hassall, C. J. S., Aimi, J., Taylor, K. M., Nakazato, H., Dixon, J. E., Burnstock, G., and Polak, J. M. (1987). Localization of atrial natriuretic peptide mRNA and immunoreactivity in the rat heart and human atrial appendage. *Proc. Natl Acad. Sci. USA*, **84**, 6760–4.

Harper, M. E. and Saunders, G. F. (1981). Localisation of single copy DNA sequences on G-banded human chromosomes by *in situ* hybridisation. *Chromosoma (Berlin)*, **83**, 431–9.

Herrington, C. S., Burns, J., Graham, A. K., Evans, M., and McGee, J. O'D. (1989*a*). Interphase cytogenetics using biotin and digoxigenin-labelled probes 1: relative sensitivity of both reporter molecules for HPV 16 detection in CaSki cells. *J. Clin. Pathol.*, **42**, 592–600.

Herrington, C. S., Burns, J., Bhatt, B., Graham, A. K., and McGee, J. O'D. (1989*b*). Interphase cytogenetics II: Simultaneous differential detection of human and HPV nucleic acids in individual nuclei. *J. Clin. Pathol.* **42**, 601–6.

Herrington, C. S., Graham, A. K., Burns, J., and McGee, J. O'D. (1990*a*). Stringency conditions for detection of closely homologous DNA sequences by non isotopic *in situ* hybridisation (NISH): HPV type discrimination using biotin and digoxigenin labelled probes. *Histochem. J.* (In press.)

Herrington, C. S., Graham, A. K., Flannery, D., and McGee, J. O'D. (1990*b*). The investigation of double infection of condylomata acuminata with HPV type 6 and 11 by nonisotopic *in situ* hybridisation. (In preparation.)

Julien, C., Bazin, A., Guyot, B., Forester, F., and Daffos, F. (1986). Rapid prenatal diagnosis of Down's syndrome with *in situ* hybridisation of fluorescent probes. *Lancet*, 863–4.

Landegent, J. E., Jansen in de Wal, N., van Ommen, G.-J. B., Baas, F., de Vijlder, J. J. M., van Duijn, P., and van der Ploeg, M. (1985). Chromosomal localisation of a unique gene by non-autoradiographic *in situ* hybridisation. *Nature*, **317**, 175–7.

Lyttleton, N. P. A., Browett, P. J., Brenner, M. K., Cordingley, F. T., Kohlman, J., McGee, J. O'D., Hamiton-Dutoit, S., Prentice, H. G., and Hoffbrand, A. V. (1988). Prolonged remission of Epstein-Barr virus associated lymphoma secondary to T cell depleted bone marrow transplantation. *Bone Transplant.*, **3**, 641–6.

Nicholls, R. D., Jonasson, J. A., and McGee, J. O'D. *et al.* (1987). High resolution gene mapping of the human globin locus. *J. Med. Genet.*, **24**, 39–46.

Pardue, M. L. (1985). *In situ* hybridisation. In *Nucleic acid hybridisation: a practical approach*, (ed. B. D. Hanes and S. J. Higgins). IRL Press, Oxford.

Reittie, J. E., Poulter, L. W., Prentice, H. G., Burns, J., Drexler, H. G., Balfour, B., Clarke, J., Hoffbrand, A. V., McGee, J. O'D., and Brenner, M. K. (1988). Differential recovery of phenotypically and functionally distinct circulating antigen presenting cells after allogeneic marrow transplantation. *Transplantation*, **45**, 1084–91.

Seabright, M. (1972). The use of proteolytic enzymes for the mapping of structural rearrangements in the chromosomes of man. *Chromosoma*, **36**, 204–10.

Shivers, B. D., Schachter, B. S., and Pfaff, D. W. (1986). *In situ* hybridisation for the study of gene expression in the brain. *Methods Enzymol.*, **124**, 497–510.

Singer, R. H., Lawrence, J. B., and Rashtchian, R. N. (1987). Toward a rapid and sensitive *in situ* hybridisation methodology using isotopic and nonisotopic probes. In In situ *hybridization: applications to neurobiology*, (ed. K. L. Valentino, J. H. Eberwine, and J. D. Barchas). Oxford University Press, Oxford.

Steel, J. H., Hamid, Q., Van Noorden, S., Jones, P., Burrin, J., Legon, S., Bloom, S. R., and Polak, J. M. (1988). Combined use of *in situ* hybridisation and immunocytochemistry for the investigation of prolactin gene expression in immature pubertal, pregnant, lactating and ovariectomised rats. *Histochemistry*, **89**, 75–80.

Sumner, A. T., Evans, H. J., and Buckland, K. A. (1971). New technique for distinguishing between chromosomes. *Nature*, **232**, 31–2.

Tank, A. W., Lewis, E. J., Chikarashi, D. M., and Wiener, N. (1985). Elevation of RNA coding for tyrosine hydroxylase in rat adrenal gland by reserpine treatment and exposure to cold. *J. Neurochem.*, **45**, 1030–3.

Terenghi, G., Polak, J. M., Hamid, Q., O'Brien, E., Denny, P., Legon, S., Dixon, J., Minth, C. D., Palay, S. L., Yasargil, G., and Chan-Palay, V. (1987). Localization of neuropeptide Y mRNA in neurons of human cerebral cortex by means of *in situ* hybridisation with a complementary RNA probe. *Proc. Natl Acad. Sci. USA*, **84**, 7315–18.

Wahl, G. M., Stern, M., and Stark, G. R. (1979). Efficient transfer of large DNA fragments for agarose gels to diazobenzyloxymethyl-paper and rapid hybridization using dextran sulphate. *Proc. Natl Acad. Sci. USA*, **76**, 3683–7.

Zabel, B. U., Kronenberg, H. M., Bell, G. I., and Shows, T. B. (1985). Chromosome mapping of genes on the short arm of human chromosome 11: Parathyroid hormone gene is at 11P15 together with the genes for insulin, C-Harvey-ras-1 and b-hemoglobin. *Cytogenet. Cell. Genet.*, **39**, 200–5.

Appendix 10.1

Materials

Tissue culture

1. Complete medium
   RPMI 76.00 ml (Gibco, Cat. No. 041-1875M)
   Fetal bovine serum 20.00 ml (NBL, Cat. No. S101)
   Phytohaemagglutinin (M FORM) 2.00 ml (Gibco, Cat. No. 061-00576C)
   Antibiotic/antimycotic solution 1.00 ml (Gibco, Cat. No. 043-05240D) (100×)
   L-Glutamine (100×) 1.00 ml (Gibco, Cat. No. 043-05030H)

2. 5'-Bromo-2-deoxyuridine (BUdR) (Sigma, Cat. No. B5002)
   Dissolve 10 mg/ml BUdR in distilled water. Sterilize by filtration with 0.2 μm filter. Store aliquots at -20°C. Handle BUdR in subdued light.

3. Thymidine (Thd) (Sigma, Cat. No. T9250)
   Dissolve 250 μg/ml in distilled water. Filter and store aliquots at -20°C.

4. Colcemid (Sigma, Cat. No. D7385)
   Dissolve 5.0 μg in distilled water. Filter and store aliquots at -20°C.

5. Potassium chloride (KCl) (Sigma, Cat. No. P9541)
   Dissolve 560 mg KCl (0.75 M) in 100 ml distilled water. Prepare fresh and keep at 37 °C.

6. Fixative
   Mix 3 parts methanol with 1 part glacial acetic acid. Prepare fresh and keep on ice.

7. Nick translation kit (Amersham, Cat. No. N5500; BRL, Cat. No. 8160SB)

8. Biotin-7-dATP (BRL, Cat. No. 9509SA)

9. Glycogen (BCL, Cat. No. 901393)

10. Ribonuclease A (Sigma, Cat. No. R9005)
    Dissolve 10 mg/ml pancreatic RNase A in 20 mM sodium acetate (pH 5.0). Place in boiling water for 10 min. Store aliquots at -20 °C. For working solution, dilute 1 in 100 with 2 × SSC.

11. 20 × SSC
    Sodium chloride 175 g (BDH, Cat. No. 10241)
    Trisodium citrate 88 g (BDH, Cat. No. 10242)
    Dissolve in 1 litre of distilled water, adjust pH to 7.4, and autoclave.

12. Hybridization mixture (HM mix)
    (a) Deionized formamide (Sigma, Cat. No. F7503)
        Mix 5 g mixed bed ion-exchange resin (Bio-Rad AG501-x8, 20–50 mesh) to 50 ml formamide. Stir for 45 min. Filter and store at -20°C.
    (b) 100 × Denhardt's solution
        Ficoll 0.2 g (Pharmacia, Cat. No. 17-0400-01)
        Polyvinyl pyrrolidone 0.2 g (Sigma, P5288)
        Bovine serum albumin 0.2 g (Sigma, Cat. No. B2518)
        Sterile distilled water to 10 ml. Store aliquots at -20°C.
    (c) Dextra sulphate (Sigma, Cat. No. D8906)
        Make 50% (w/v) solution in sterile distilled water.
    (d) Human or salmon sperm DNA (Sigma, Cat. No. D7011 or D1626)
        Dissolve 10 mg/ml in distilled water. Stir for 2–4 h. Shear DNA by passing it several times through 18-gauge needle or by sonicating. Place in boiling water for 10 min. Store aliquots at -20°C.

    HM mix (store at 4 °C)
    Deionized formamide 6.0 ml
    100 × Denhardt's solution 1.0 ml
    50% Dextran sulphate 2.0 ml
    20 × SSC 1.0 ml

13. PBT solution
    This consists of 5% bovine serum albumin (BSA) and 0.1% Triton X-100 (Sigma) in phosphate buffered saline. BSA solutions should be filtered through Whatman 3 mm before use.

14. Primary antibody (Enzo Biochem Inc., N.Y., Cat. No. EBP-861-2)
    This is a polyclonal antibody to biotin raised in rabbit (rabbit α-biotin). Monoclonal antibiotin is also now available from Dako, UK. Dilute antibiotin 1 in 100 with PBT.

15. Second antibody (Janssen, Cat. No. EMGARG15)
    Goat anti-rabbit 1gG conjugated to 10–20 nm gold. Dilute 1 in 50 with PBT.
16. Silver solution (Janssen, Cat. No. 30.115.45)
17. Hoechst 33258 (Sigma, Cat. No. B2883.
    $100 \times$ stock solution is prepared by dissolving 5 mg Hoechst dye in 10 ml methanol. For working solution the stock is diluted 1 in 100 with $2 \times SSC$.

Preparation of whole blood cultures

1. Collect blood in preservative-free heparin (20 U/ml).
2. Add 0.8 ml whole blood to each culture bottle containing 10 ml of complete medium. Incubate at 37 °C for 72 h.
3. Add 100 μl BUdR to each culture and incubate at 37 °C for a further 16–17 h.
4. Centrifuge at 1200 rpm for 8 min. Discard supernatant and resuspend pellet in 10 ml unsupplemented RPMI 1640 medium.
5. Repeat step 4.
6. Resuspend cells in 10 ml fresh complete medium containing 2.5 μg/ml thymidine. Return cultures to 37 °C for 6–7 h.
7. Colcemid may be added 15–30 min prior to harvesting the cells, although this is not absolutely essential when cultures are synchronized by either BUdR or methotrexate.
8. Centrifuge cultures at 1200 rpm, discard the supernatant and resuspend the pellet in 1 ml of 0.75 M (0.56%) KCl pre-warmed to 37 °C with a glass Pasteur pipette. Bring the volume to 10 ml with 0.75 M KCl and incubate at 37 °C for 10 min.
9. Spin cells as before and resuspend the pellet in fresh chilled fixative. Leave on ice for at least 20 min.
10. Repeat step 9 several times or until the cell suspension appears clean.
11. The cells may be left in the fixative indefinitely at − 20 °C.
12. Alternatively, centrifuge and resuspend pellet in 0.5–1.0 ml fresh fixative and keep on ice.

Preparation of chromosomal spreads

1. Soak slides in Decon (or any other detergent such as Lipsol) overnight. Rinse in running tap water followed by a few rinses in distilled water. Dry at 60–75 °C.
2. Incubate slides in a 2% solution of silane in acetone at room temperature for 60 min (Burns *et al.* 1987).

3. Rinse in acetone for 5 min, followed by tap water.

4. Dry at 60–75 °C.

5. Place 10 μl cell suspension on each glass slide and allow to air-dry. Allow the preparations to 'age' for at least 24 h. Slides may be stored for up to 8 weeks.

Removal of endogenous RNA

1. Place 100 μl of RNase solution on each preparation, under a coverslip and incubate (in Terasaki plates containing a few drops of $2 \times$ SSC) at 37 °C for 60 min.

2. Allow the coverslips to float off in $2 \times$ SSC and wash in two changes of $2 \times$ SSC.

3. Dehydrate through a series of 10%, 50%, 70%, 90%, and absolute ethanol.

4. Air dry; (slides may be stored in a desiccator if they are not going to be used immediately).

Labelling DNA probes

Label 1.0 μg of probe by nick translation with biotin-11-dUTP or biotin-7-dATP using the standard nick-translation kit (BRL, UK), in the following way.

dNTP	5.0 μl
Bio-7-dATP	2.5 μl
Plasmid 1.0 μg	(1.0 μg)
Enzyme	5.0 μl
10^{-2} M MgCl$_2$ to 50 μl	

1. Incubate at 15 °C for 90 min.

2. Add 5 μl 0.2 M EDTA to stop reaction.

3. Heat-inactivate at 75 °C for 5 min.

4. Add 1 μl of mussel glycogen (20 mg/ml), as carrier.

5. Add 1/10 volume sodium acetate (3M) pH 6.0.

6. Add 2 volumes of absolute ethanol.

7. Incubate at -70 °C overnight or on dry ice for 2 h.

8. Centrifuge in eppendorf for 8–10 min.

9. Discard supernatant and wash pellet four times with 80% ethanol.

10. Lyophilize for 10 min.

11. Dissolve in 50 μl TE or distilled water.

Denaturation/hybridization

1. Add 20–100 ng of biotin-labelled probe to hybridization mixture. Add

5 μl of 20 mg/ml sonicated human or salmon sperm DNA. The total volume should be 30–40 μl. Mix and spin in eppendorf centrifuge for 5 s.

2. Apply the mixture to the chromosomal preparations, under a coverslip. Seal the edges of the coverslip with rubber cement. (Dunlop tyre glue).

3. Place the preparations in Terasaki plates and incubate at 75 °C for 7–8 min.

4. Incubate at 37 °C for 16 h.

Removal of non-specifically bound probe (stringency washes)

1. Remove coverslips and incubate slides in 50% formamide in 2 × SSC (pH 7.2) for 20 min at 42 °C.

2. Rinse in 2 × SSC for 20 min at 42 °C.

3. Rinse in 1 × SSC at room temperature for 20 min.

4. Rinse in 0.1 × SSC at room temperature for 20 min.

Immunocytochemical detection of bound probe

1. Incubate slides in PBT for 15 min.

2. Dry the back and edges of the slides and place 100 μl of primary antibody on each slide and incubate at 37 °C for 45–60 min.

3. Rinse briefly in PBT and dry the back and edges of slides.

4. Place 100 μl of second antibody on each slide and incubate at 37 °C for 45–60 min.

5. Rinse in PBS for 3 × 5 min.

6. Rinse in distilled water for 3 × 5 min.

7. Place a few drops of silver solution on each slide and monitor under light microscope. Development of silver dots takes 5–18 min.

8. Wash in distilled water.

Replication banding

9. Stain slides with 5 μg/ml Hoechst 33258 in 2 × SSC for 20 min. Keep the preparations in dark.

10. Rinse in distilled water, mount in 2 × SSC and seal the coverslips with rubber cement.

11. Expose the slides to UV light for 1–16 h.

12. Rinse in distilled water and stain with 10% Giemsa made in phosphate buffer (pH 6.8–7.2) for 10 min.

13. Rinse in distilled water, air-dry, and mount in DPX.

11

Interphase cytogenetics of solid tumours

A. H. N. HOPMAN, F. C. S. RAMAEKERS, AND G. P. VOOIJS

Introduction

Quantitative and structural aberrations in the genomic content of malignant or premalignant lesions are in some cases correlated with the prognosis of the disease. Since such genetic changes may be central to the initiation and progression of neoplasms, techniques have been developed for their detection and characterization.

In particular, flow cytometric (FCM) techniques and karyotyping have been used for such analyses. In FCM assays the total DNA content of a tumour cell population is quantitated. This technique, however, does not give information with respect to specific chromosome aberrations and has limitations in the detection of minor quantitative DNA changes. Karyotyping, on the other hand, facilitates the identification of small deviations in chromosome content and chromosome structure. However, chromosome analysis of cancer cells by karyotyping (metaphase cytogenetics) is often only possible after tissue culturing. This method may result in a selective growth of cells with the highest mitotic index and loss of chromosomal material. Furthermore, such analyses are often hampered by the small number of recognizable metaphases, the lack of chromosome spreading, poor banding quality, and a condensed or fuzzy appearance of the chromosomes.

To overcome these limitations of FCM and karyotyping assays, *in situ* hybridization (ISH) procedures using chromosome-specific DNA probes have been developed (Rappold *et al.* 1984; Manuelidis 1985; Burns *et al.* 1985; Schardin *et al.* 1985; Cremer *et al.* 1986; Trask *et al.* 1988). These studies demonstrate that chromosomes occupy discrete areas within interphase nuclei. The ISH method enables the detection of numerical and structural chromosome aberrations in interphase nuclei (Cremer *et al.* 1988*a,b*; Devilee *et al.* 1988; Hopman *et al.* 1988*b*, 1989; Lichter *et al.* 1988, 1989; Pinkel *et al.* 1988; Dekken and Bauman 1988; Emmerich *et al.* 1989; Nederlof *et al.* 1989*a,b*) and therefore, this technique has been referred to as 'interphase cytogenetics' (for review see Herrington and McGee 1990).

Nucleic acid probes used in ISH for localizing specific nucleotide sequences in morphologically preserved cells, chromosome spreads, or tissue sections can be labelled by enzymatic incorporation of radioactive precursors (Chapter 3). Such radioactively labelled probes enable, e.g., the detection of mRNAs (Angerer *et al.* 1985; Coghlan *et al.* 1985), single-copy genes in human metaphase spreads (Gerhard *et al.* 1981; Harper *et al.* 1981), viral genomes (Haase *et al.* 1985), and viral RNA within single cells (Harper *et al.* 1986). In spite of the high sensitivity of ISH with radio-actively labelled probes (RISH), this method has several disadvantages, such as the relatively long exposure times for detection, biological hazard, and the inconvenience of disposal. Furthermore, radioactive labelling and visualization by autoradiography has the disadvantage of the limited spatial resolution, which provides a severe limitation for interphase cytogenetics because of the small size of the nucleus. Modification of nucleic-acid probes with haptens or other labels, followed by non-isotopic detection by, e.g., fluorochromes, enzymes, or colloidal gold particles, overcome these problems (for review see Bauman *et al.* 1984; Hopman *et al.* 1988*a*; Raap *et al.* 1989; McGee *et al.* 1987; Matthews and Kricka 1988; van der Ploeg and Raap 1988). Furthermore, non-isotopic ISH (NISH) enables the detection of two, three, or even more different DNA targets by using differently labelled probes followed by different immunocytochemical detection systems (Hopman *et al.* 1986, 1988*a,b*; Singer *et al.* 1987; Cremer *et al.* 1988*a*; Emmerich *et al.* 1989; Herrington *et al.* 1989*a,b*; Nederlof *et al.* 1989*b,c*). NISH has been shown to be as sensitive as RISH in the detection of unique sequences on metaphase chromosomes and interphase cells (Chapter 10), single viral sequences and viral RNAs in interphase cells in culture (Singer *et al.* 1987; Lawrence *et al.* 1989), and for viral detection in archival biopsies (Burns *et al.* 1987; see also Chapter 12).

Technical aspects of *in situ* hybridization

The hybridization and detection protocol, which we apply to single cell suspensions, paraffin sections and frozen sections of solid tumours, has the following general outline:

(1) probe modification for non-isotopic detection;

(2) fixation of the biological material, slide preparation, and pretreatments of tissue material on the slides;

(3) hybridization of the modified probe with denatured target DNA, washing, and immunocytochemical detection.

DNA probes for interphase cytogenetics

To detect numerical and structural chromosome aberrations several approaches can be followed.

Probes recognizing highly repetitive sequences (Singer 1982; Willard and Waye 1987)

DNA probes applicable in interphase cytogenetics for non-isotopic labelling of specific chromosomes in the (peri)centromeric region are now available for human chromosomes 1 (Cooke and Hindley 1979; Waye *et al.* 1987*a*), 7 (Waye *et al.* 1987*b*), 8 (Dolon *et al.* 1986), 9 (Moyzis *et al.* 1987), 10 (Devilee *et al.* 1988), 15 (Higgins *et al.* 1985), 16 (Moyzis *et al.* 1987), 17 (Waye and Willard 1986), 18 (Devilee *et al.* 1986), X (Willard *et al.* 1983), and Y (Cooke *et al.* 1982). In most instances, probes of the satellite or alphoid family have been used. The targets of these probes occur in several hundreds up to several thousands of higher-order repeats, resulting in DNA targets up to several thousand kilo-base pairs (kbp) localized in the compact centromere regions of the individual chromosomes. For example, the DNA probe for chromosome 1 (pUC 1.77) recognizes a tandem repeat of 1.77 kb in the (peri)centromeric region 1q12 (Cooke and Hindley 1979). Since these chromosome regions appear as distinct DNA clusters within the interphase nucleus, and their number is constant in normal cells, these probes allow relatively reproducible non-isotopic detection (Cremer *et al.* 1988*a*; Devilee *et al.* 1988; Hopman *et al.* 1988*b*, 1989; Nederlof *et al.* 1989*b*; Emmerich *et al.* 1989). Recently, an additional probe has been published that detects a specific repetitive target sequences in the telomere region (1p36) of chromosome 1 (Buroker *et al.* 1987; Dekken *et al.* 1988; Dekken and Bauman 1988).

Probes recognizing total chromosomes or parts of chromosomes

For this technique (also referred to as chromosomal *in situ* suppression hybridization or competition hybridization) genomic DNA libraries, originating from flow cytometrically sorted chromosomes or DNA isolated from cell lines are used (Cremer *et al.* 1988; Lichter *et al.* 1988; Pinkel *et al.* 1988; Kievits *et al.* 1989; Landegent *et al.* 1987). In order to obtain chromosome-specific signals, hybridization of the library DNA with ubiquitous sequences in the other chromosomes is suppressed by the addition of total genomic human DNA. As a result, a more or less homogeneous hybridization signal on individual human chromosomes from pter to qter can be obtained. Recombinant libraries for chromosomes 1, 4, 7, 8, 13, 14, 18, 20, 21, 22, and X have so far been applied in this way. In another approach, composite probes compiled from several sequences cloned from the desired target region are used. For this purpose a panel of plasmid clones derived from cosmids that have been mapped to specific chromo-somes, or a library of cosmids or phages, were used for hybridization (Langedent *et al.* 1987; Pinkel *et al.* 1988; Lichter *et al.* 1989).

Labelling of probes and immunocytochemical detection

Several methods for non-isotopic detection are described that are based on the introduction of a reporter molecule into the DNA probe that can then be detected immunocytochemically after hybridization (Bauman *et al.* 1984; Hopman *et al.* 1988*a*; Raap *et al.* 1989; Matthews and Kricka 1988; van der Ploeg and Raap, 1988). The labelling procedures include: enzymatic incorporation of biotin (Brigati *et al.* 1982; Langer *et al.* 1982) or chemical introduction of the biotin molecule into the probe (Forster *et al.* 1985); enzymatic incorporation of BrdUTP (Frommer *et al.* 1988); chemical modification by introduction of an acetylaminofluorene (AAF) group (Landegent *et al.* 1984, 1985; Tchen *et al.* 1984); labelling with mercury ions, to which a thiol-hapten molecule is coupled after hybridiz-ation (Hopman *et al.* 1986*a,b,d*); chemical sulphonation (Pezzella *et al.* 1987); and labelling with digoxigenin (Herrington *et al.* 1989*a,b*). Probe preparation is discussed in Chapters 1, 3, and 4.

In most cases the haptens, or the modified probe, are detected by affinity cytochemical techniques. In the case of biotin, the reporter molecule can also be detected with labelled streptavidin. The following systems can be used. Indirect immunocytochemical procedures using fluorochromes such as FITC (Cremer *et al.* 1988*a,b*; Hopman *et al.* 1988*b*; Trask *et al.* 1988; Emmerich *et al.* 1989), TRITC (Cremer *et al.* 1988*a*; Devilee *et al.* 1988; Emmerich *et al.* 1989; Hopman *et al.* 1989), or AMCA (aminomethyl-coumarin acetic acid) (Nederlof *et al.* 1989*a,c*). Indirect immunocyto-chemical procedures using enzymes such as peroxidase with or without amplification (Manuelidis and Ward 1985; Burns *et al.* 1985; Cremer *et al.* 1986; Hopman *et al.* 1986*b,c*; Raap *et al.* 1986; Naoumov *et al.* 1988; Emmerich *et al.* 1989*a,b*), alkaline phosphatase (Hopman *et al.* 1986*c*; Burns *et al.* 1987; Emmerich *et al.* 1989*b*), or avidin/biotinylated–anti-avidin system (Lichter *et al.* 1988; Pinkel *et al.* 1986, 1988; Trask *et al.* 1988). For digoxigenin, indirect systems employing primary antibodies (monoclonal or polyclonal) against digoxigenin, and second antibodies linked to alkaline phosphatase, have been successful (Herrington *et al.* 1989*a,b*). Amplification of digoxigenin detection systems is described in Chapter 12.

Multiple-target *in situ* hybridization

Simultaneous detection of several probes in one cell, chromosome spreads or tissue sections can be done with probes bearing different haptens. These are visualized with different distinguishable affinity systems. The resolution between two different NISH signals using fluorochromes, e.g. fluorescein (FIFC) in combination with rhodamine (TRITC) is excellent. Recently, even three DNA probes have been simultaneously detected with FITC, TRITC and AMCA as fluorescent labels (Nederlof *et al.* 1989). Multiple-target NISH can be used to demonstrate numerical as well as structural chromosome aberrations (Hopman *et al.* 1988, 1989; Cremer *et al.* 1988*a,b*; Nederlof *et al.* 1989). Also peroxidase and alkaline phosphatase have been applied in precipitation reactions to detect two different nucleic acids by ordinary light microscopy (Emmerich *et al.* 1989*b*; Hopman *et al.* 1986; Herrington *et al.* 1989*a,b*).

Tumour processing for *in situ* hybridization

Fresh tumour material obtained after surgery or by fine-needle aspiration biopsies was divided for the different diagnostic approaches depicted in Fig. 11.1. The individual tumour blocks were treated as follows.

1. Storage in liquid nitrogen and subsequent cryosectioning.

2. Fixation in 4% formaldehyde buffered in phosphate-buffered saline (PBS) for different time periods, up to 24 h, and embedding in paraffin. These tissue blocks were used for routine histopathological diagnosis, immunocytochemistry and NISH on 4–7 μm thick sections.

3. Preparation of single cell suspensions for FCM and NISH (Smeets *et al.* 1987; Hopman *et al.* 1988*b*; Feitz *et al.* 1985). In brief, the tissues were mechanically disaggregated by scraping and cutting in a Petri dish and filtered through a 100-μm nylon filter (Ortho Diagnostic Systems, Beerse, Belgium). The filtered cell suspensions were fixed in 70%

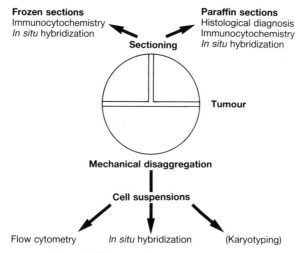

Fig. 11.1. Flow chart for tumour processing.

ethanol at −20 °C, and stored for some months to several years at −30 °C.

Slide preparation and pretreatment of tissues

Isolated cells or tissue sections may be lost during the ISH procedure when they are attached to uncoated glass slides. This can be avoided by coating with poly-L-lysine, glutaraldehyde activation of gelatin–chromealum-coated slides (Raap *et al.* 1988) and coating of the slide with aminoalkylsilane (Tourtelotte 1986; Burns *et al.* 1987; Van Prooyen-Knegt *et al.* 1982).

In order to detect specific nucleic acid sequences, DNA probes and their detectors (e.g. antibodies, etc.) have to penetrate complex and dense biological structures to reach their targets. For this reason, the biological material has to be pretreated to increase accessibility of these DNA targets.

Pretreatment of tumour cell suspensions

Cells that were isolated by mechanical disaggregation show a high auto-fluorescent background because of remaining cytoplasm resulting in reduction in the fluorescence intensity of the NISH signals where the nucleus was covered by cytoplasm. To remove the cytoplasm and part of the nuclear proteins, two procedures were followed.

Postfixation in methanol/acetic acid

Shortly before use, cells fixed in ethanol were postfixed with freshly prepared methanol/acetic acid ($3/1$, v/v for 4×5 min at 0 °C). Two or three drops of the cell suspension were placed onto the coated slides and air-dried. Optionally, the cytoplasm was partly removed by dipping the slides in 70% acetic acid for 10–60 s, washed twice with distilled water, dehydrated in 100% ethanol, and air dried.

Proteolytic digestion with pepsin

After dropping 5 μl of the cell suspension onto the slides, they were air dried for 30 min (optional at 80 °C). Pepsin, from porcine stomach mucosa (2500–3500 units per mg protein, Sigma, USA), was applied at a concentration of 50–400 μg/ml in 0.01 M HCl for 10 min at 37 °C. After washes in H_2O, and PBS, followed by fixation in 4% formaldehyde (prepared from paraformaldehyde) in PBS for 5 min at 0 °C, the slides were dehydrated, air dried, and heated at 80 °C for 30 min. Formaldehyde fixation was found to be essential for retention of the cells on the glass slides as well as for better preservation of the nuclear morphology and general DNA staining.

This second approach was most reproducible for protein removal or exposure of the DNA target and thus for penetration of the probe. Furthermore, it gave the best morphology and the most discrete NISH signals with high fluorescence intensity. Cells fixed with methanol/acetic acid were often flat compared with the pepsin-treated cells. Microscopic examination of the latter cells, however, was difficult and evaluation of NISH signals required focusing at different levels of the nucleus. In general, for pepsin treatment, cells with a higher DNA content and larger nuclei need lower enzyme concentrations than cells with smaller nuclei to obtain optimal hybridization signals. For practical reasons, we normally used the highest concentration of pepsin that still gave acceptable nuclear morphology.

Both protocols were tested on ethanol-fixed human lymphocytes, the bladder carcinoma cell line T24 and the leukaemic tumour cell line Molt-4; these showed two, three, and four NISH spots per nucleus, respectively for the chromosome 1 probe. These cells were used to develop the proper cell-preparation techniques, to optimize the specificity and intensity of the technique to obtain maximal fluorescence signals, and to test the conditions influencing detection of numerical chromosome aberrations. The results of NISH on the Molt-4 cells are shown (Fig. 11.2) to demonstrate the importance of the pepsin concentration in pretreatment for the detection of chromosomal copy number.

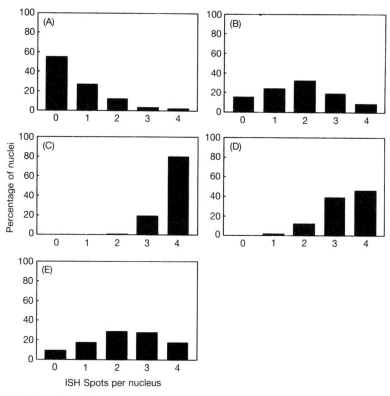

Fig. 11.2. Effect of pepsin concentration on NISH detection on tetrasomy for chromosome 1 in Molt-4 cells. Cells, on poly-L-lysine-coated slides, were enzymatically treated with pepsin to improve probe penetration. Concentrations were (A) 0 μg/ml, (B) 5 μg/ml, (C) 20–200 μg/ml, (D) 800 μg/ml, (E) 1600 μg/ml.

The frequency distribution of the NISH spots as a function of pepsin concentration shows that the concentration of pepsin is critical. An underestimate of the copy number can easily be made when the pepsin concentration is too low, while large heterogeneity in the copy number and poor morphology is a result of overdigestion. However, within a certain range of pepsin concentrations the copy number detected in most of the cells equals the real copy number. It should be stated, however, that cells with lower copy numbers can always be detected as a result of inefficient hybridization and colocalization of spots. For these reasons, about 10 per cent of such (artificial) numerical aberrations are tolerated. An overestimation of signal numbers can be made when the morphology of the cells is disrupted.

Pretreatment of sections

Different proteolytic enzymes including pepsin, proteinase K, and pronases were used in different NISH protocols (Van Prooijen-Knegt *et al.* 1982; Burns *et al.* 1985; Hopman *et al.* 1987; Raap *et al.* 1986, 1988; Naoumov *et al.* 1988). We found that the best assessment of NISH signals was obtained with pepsin digestion in HCl (Burns *et al.* 1987). The following protocols resulted in specific hybridization signals with minimal non-specific binding in about 60 per cent of cases.

Frozen sections

Sections 5 μm thick on coated slides were air dried and fixed in 4% formaldehyde in PBS at 0 °C for 15 min. After fixation, and washing with PBS and 0.01 N HCl, the sections were digested with pepsin at a concentration of 20–100 μg/ml in 0.01 N HCl for 30 min at 37 °C. After two washes in PBS for 5 min, postfixation in 4% formaldehyde for 15 min at 0 °C, slides were washed twice in PBS, dehydrated through graded alcohols, and air dried. Before NISH, the slides were heated for 30 min at 80 °C.

Paraffin sections

Sections 5 μm thick were floated in warm demineralized water (40 °C) and mounted on coated slides. Sections were air dried, heated for 1–16 h at 56 °C, dewaxed in xylene (3 × 10 min), rinsed twice in methanol for 5 min, and endogenous peroxidase activity was blocked in 1% H_2O_2 in methanol for 30 min. Subsequently, the slides were rinsed in methanol, air dried, and digested with pepsin (4 mg/ml in 0.2 N HCl) for 30 min at 37 °C. After digestion, slides were washed for 5 min in demineralized water, fixed in 70% ethanol, and washed again in PBS. Before hybridization the slides were incubated for 15 min in 1% glycine in PBS, washed with PBS and dehydrated through graded alcohols and air dried. Optionally, slides were heated for 30 min at 80 °C.

Hybridization

Probe modification

Single-target NISH was performed with biotinylated probes. Biotinylation of probes was performed using bio-7-dUTP (BRL, USA) in a nick-translation reaction (see Chapter 1). The fragment length of the biotinylated probes was about 200–400 bases. This was achieved by varying the DNase concentration during the reaction. In double-target NISH biotinylated probes were combined with mercurated probes. Prior to mercuration of the probes, DNA was sonicated to obtain fragments of about 400 bases. Mercuration was performed using mercury(II)acetate and the DNA isolated as a cyanide complex. After purification, the nucleic acids were dissolved to a concentration of 2–6 ng/μl in the hybridization mixture.

Hybridization and detection

The probes for chromosome 1 (1c) (Cooke and Hindley 1979) and chromosome 18 (18c) (Devilee *et al.* 1986) not only showed binding to chromosomes 1 or 18, but also to chromosomes 9, 16 and 13, 21 respectively, when low-stringency conditions were applied (e.g. 50% formamide in 2 × SSC at 37 °C). To avoid hybridization to these minor binding sites, the hybridization mix contained 60% (v/v) formamide, 2 × SSC, pH 5.0, and salmon sperm DNA, 1 μg/μl. For double-target NISH, 1 mM KCN was added to the hybridization buffer. Biotinylated probes were efficiently hybridized in double-target NISH without interference by cyanide ions, which are essential for hybridization of mercurated probes and without interference of the mercurated probes (Hopman *et al.* 1986*a,b,d*). The hybridization mixture was added to dry slides (5 μl per 18 × 18 mm coverslip), sealed with rubber cement under a coverslip, and denatured together with the target DNAs at 80 °C on a heating plate. Preparations for NISH on single-cell suspensions were denatured for 2.5 min, while sections were heated 5–10 min. Incubations with the specific DNA probes were performed overnight at 37 °C. Post-hybridization washings for double-target NISH were done in 60% formamide, 2 × SSC, pH 5.0, 1 mM KCN, for 15 min; once in 2 × SSC, pH 5.0, for 15 min; and twice in 3 × SSC, pH 5.0, containing 1 mM EDTA and 0.05% Tween 20 (peroxide-free, Pierce, USA) for 5 min. Immunocytochemical detection of the mercurated probe was performed by subsequent incubation with the mercury-binding ligand trinitrophenyl-lys-lys-NH-CH$_2$-CH$_2$-SH (Hopman *et al.* 1986*a,b*) at a concentration of about 1 μM in 1 mM EDTA, 3 × SSC, pH 5.0. Thereafter, sheep anti-trinitrophenyl serum, diluted 1:200 in PBS containing 2% normal rabbit serum and 0.05% Tween 20 (buffer A) was added, followed by incubation with FITC-conjugated rabbit anti-sheep serum (Nordic Immunology, Netherlands), diluted 1:80 in buffer A. Biotin labelled DNA both in single- and double-target NISH was detected by

TRITC-conjugated avidin (Sigma, USA) diluted 1:1000 in buffer A. After each incubation step, the slides were washed three times in PBS for 10 min. For the fluorescent approach, the slides were dehydrated and mounted in PBS/glycerol (1:9 = v/v) containing 2.3% 1,4-di-azobicyclo-(2,2,2,)-octane (Sigma, USA). In double-target NISH, the FITC and TRITC-conjugated reagents were incubated simultaneously. Microphotographs were taken with a Leitz Dialux 20 EB microscope, equipped for FITC, TRITC, and DAPI fluorescence using a 3M colour slide film (640-T, 3200 K). Exposure times were 2–3 min for FITC fluorescence, approximately 40 s for TRITC fluorescence, and 1 s for DAPI fluorescence.

Sections incubated for single-target NISH were washed as described above and the biotinylated probes were detected indirectly with mono-clonal anti-biotin (DAKOPATTS, Denmark), followed by peroxidase-conjugated rabbit anti-mouse antibodies and a DAB precipitation reaction.

Protocols for the use of chromosome 15 (Higgins *et al.* 1985) and 17 (Wayne and Willard 1986) probes are virtually identical to the protocols for 1c and 18c described above.

Criteria for evaluating NISH preparations

The following criteria have to be applied to allow a proper evaluation of *in situ* hybridization signals on isolated tumour cells.

- Nuclei should not overlap.
- Cells should not be asymmetrically covered by cytoplasm.
- NISH signals should have more or less the same homogeneous fluorescence intensity.
- Minor hybridization spots (Fig. 11.3(G)), which can be recognized by a lower intensity, should not be counted.
- Fluorescent spots (Fig. 11.3(A)) or patches of fluorescence (Fig. 11.3(B,F)) may only be included when the signals are completely separated from each other.
- Spots in a paired arrangement (split spots; Fig. 11.3(C)), close to each other are counted as one chromosome complement.

When the cell population with negative nuclei exceeded 15 per cent the procedure should be regarded as suboptimal; this means that the chromosome ploidy estimation may be incorrect. When these criteria were applied, interobserver variations were minimal.

Applications

The NISH methods described above have been applied to several types of malignancies including breast cancer (Devilee *et al.* 1988), gynaecological tumours (Nederlof *et al.* 1989), neurological tumours (Cremer *et al.* 1988*a,b*), testicular tumours, carcinoma *in situ* of the testis (Giwercman *et*

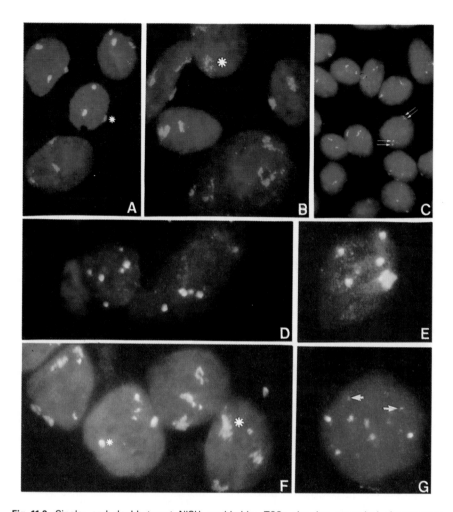

Fig. 11.3. Single- and double-target NISH on bladder TCCs showing numerical chromosome aberrations. (A–D) Single-target NISH on isolated TCC (DI = 1.0) cells with the probe for chromosome 1, showing trisomy for chromosome 1 and a distinct manifestation of spots (small asterisk). (B) Blurred appearance of NISH signals for chromosome 1 (large asterisk). (C) Split spots for chromosome 1 targets (arrows). (D–G) Double-target NISH on isolated bladder TCC cells with probes for chromosomes 1 and 18. (D) Trisomy for both probe chromosome 1 (TRITC) and chromosome 18 (FITC) in a TCC with a DNA index of 1.5. (E) Similar to (D), but more than 5 copies of chromosome 1 were detected, while the signals for chromosome 18 could not be analysed. (F) Diploid TCC showing trisomy for chromosome 1 and disomy for chromosome 18. (G) Tetraploid TCC with disomy for chromosome 1 and tetrasomy for chromosome 18. Arrows indicate examples of minor binding sites. Microphotographs were taken by (C) single, (A, B) double, and (D–G) triple exposure. In (A, B, D–G) DNA was counterstaining with DAPI; in (C) DNA was counterstaining with propidium iodide.

al. 1989; Emmerich *et al.* 1989), and urinary bladder cancer (Hopman *et al.* 1988*b*, 1989). The results on the latter two tumour types will be illustrated below.

Detection of numerical chromosome aberrations
Cell suspensions

In order to detect numerical chromosome aberrations in transitional cell carcinomas (TCCs), single-cell suspensions of these tumours were hybridized with centromere-specific DNA probes for chromosomes 1, 15, 17, and 18.

NISH studies on bladder tumours, which had a DNA index of approximately 1.0 as measured by FCM, have shown that in 'diploid' TCCs chromosome copy numbers that deviated from two could be detected frequently. An example, is shown in Fig. 11.3(A,B), in which three spots for the chromosome 1 probe were detected. In about 30 per cent of TCCS, which showed a single DNA peak with a diploid DNA index in FCM, a significant population of cells with trisomy for chromosome 1 could be detected. FCM in these cases was unable to detect these quantitative aberrations in the genomic content. By means of double-target NISH, we were able to demonstrate that numerical differences of individual chromosomes exist within one tumour. As depicted in Fig. 11.3(F), the main copy number for chromosome 1, in the case illustrated was 3; in the same cells the main copy number for chromosome 18 was 2. NISH on tumours with higher DNA-indices (as determined by FCM), showed that the copy number for chromosome 1 in these tumours varied over a wide range. The examples in Fig. 11.3(D,E) illustrate that, in a TCC with a DNA index of 1.5, cells with varying copy numbers of chromosome 1 occur. Furthermore, Fig. 11.3(G) shows a cell with two copies of chromosome 1 in a 'tetraploid' TCC.

Double-target NISH demonstrates an even more profound heterogeneity in these malignancies. Extensive heterogeneity within these tumours is also demonstrable by hybridizing with different centromeric probes (e.g. to chromosomes 1, 15, 17, and 18). Figure 11.4 shows NISH data with all of these probes on diploid human lymphocytes, the stable triploid bladder cell line T24, and TCCs with a DNA index of 1.0 and 3.2. In lymphocytes and T24, all DNA probes showed NISH spot numbers as expected on the basis of karyotyping (data not shown); but the TCCs showed a more aberrant spot number pattern. For example, an apparently diploid TCC exhibited cell populations with disomy for all probes, but in a significant number of cells there was monosomy for chromosomes 15 and 17. In a TCC with a DNA index of 3.2, a higher copy number for chromosome 1 than for chromosomes 15, 17, and 18 was detected. Since by NISH we are generally able to analyse the total tumour cell population, we could detect minor cell

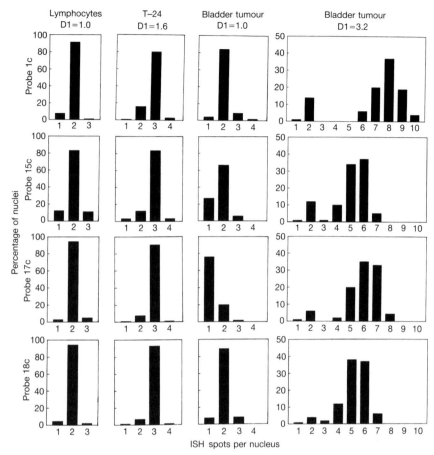

Fig. 11.4. NISH on human lymphocytes, a T24 cell culture, and two TCCs with probes for chromosomes 1, 15, 17, and 18. For evaluation, 100–200 cells were counted. Probe 1c = chromosome 1, etc.

populations with very high chromosome copy numbers. These cells may possibly originate from polyploidization.

Paraffin sections

In order to evaluate the possibility of detecting chromosome copy number in paraffin sections, we correlated copy number as detected in cells isolated from fresh tumour material with the number detected in paraffin sections of the same tumour (Fig. 11.5(B,C), 11.6(A–D)). As an example, we show a TCC with a DNA index of 1.0, in which the main population had trisomy for chromosome 1 in single-cell suspensions. The inflammatory cells in the paraffin sections of this tumour mainly showed two spots for chromosome 1 (Fig. 11.5(A)). In the tumour areas, cell populations with

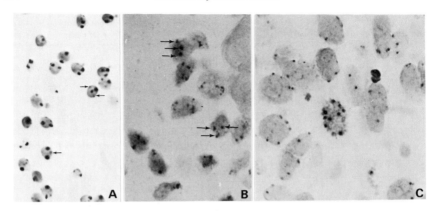

Fig. 11.5. NISH on paraffin sections with the probe for chromosome 1 on inflammatory cells in a TCC (A), and cells of a TCC with a DI of 1.0 (B), and a TCC with a DI of 3.2 (C). Arrows indicate examples of minor binding sites in (A) and NISH detection of trisomy (B).

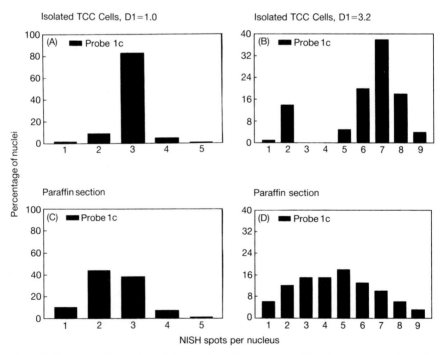

Fig. 11.6. Frequency distribution of the number of chromosome 1 NISH spots per nucleus after analysing 150–300 nuclei in cells isolated from fresh tumours (A, B) and in paraffin sections (C, D) of the same tumour specimens. (A, C) TCC with a DI of 1.0 and (B, D) TCC with a DI of 3.2.

spot numbers ranging from 1 to 4 were detected (Figs. 11.5(B), 11.6(C)). Comparison of the NISH data from isolated tumour cells and paraffin sections show that aneuploidy can easily be detected (Figs. 11.5(B), 11.6(C)). Sectioning of nuclei in paraffin sections can result in under-estimating the real copy number. This was pronounced in the TCC with a DNA index of 3.2 (Fig. 11.5(C)). In isolated cells, the average spot number ranged from 6 to 8, while in paraffin sections this range was larger and no evident major peak could be detected (compare Figs. 11.6(B) and (D)). However, in the latter aneuploid cells were evident. In this TCC, some mitoses were strongly positive for the chromosome 1 probe, frequently showing copy numbers in the range of 5–9.

Detection of structural chromosome aberrations

Structural chromosome aberrations can generally not be detected with centromere-specific probes. Since such structural changes involve trans-locations or deletions of the p or q arm (or parts of these arms), selective staining of the complete chromosome is necessary for the detection of these changes. Telomere probes may provide some help in this respect (Dekken and Bauman 1989). Recent developments, which may in future allow the detection of structural chromosome aberrations in interphase nuclei, involve the use of libraries of DNA probes in combination with competition *in situ* suppression hybridization (CISS). With this approach, a particular chromosome can be labelled as a whole and be visualized both in meta-phase spreads and in interphase nuclei (Lichter *et al.* 1988; Emmerich *et al.* 1989). Figures 11.7(A) and (B) illustrate this hybridization approach in a cell line derived from a germ-cell tumour. The three individual chromo-somes 1, next to three chromosome 1 translocations, are clearly seen in the interphase nucleus. The results with metaphase spreads (Fig. 11.7(A)) of these cells show that the probes stain the complete chromosome 1, as well as the translocated parts of chromosome 1. It is also obvious from Fig. 11.7 that the results in metaphase spreads correlate extremely well with those from interphase nuclei.

Conclusion

NISH procedures can be used for rapid and sensitive detection of numerical and structural chromosome aberrations in interphase nuclei. The applicability of these methods to interphase cytogenetics is dependent on the DNA probes that are available and the sensitivity of the NISH detection procedures. Probes recognizing highly repetitive DNA sequences in the centromeric region of the individual chromosomes are used to detect numerical chromosome aberrations. Detection of the chromosome copy number is, however, strongly influenced by the pretreatment applied to the tumour material. For example, underestimation of real chromosome copy

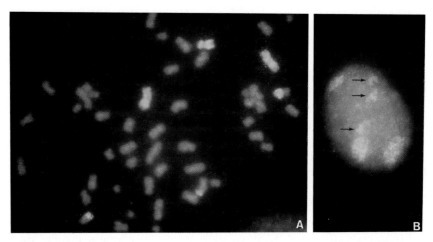

Fig. 11.7. Staining of total chromosomes by chromosomal *in situ* suppression hybridization. In this case of a human male germ-cell tumour cell line (Germa 2), a library of chromosome 1 inserts was used. (A) Metaphase spread showing two apparently normal chromosomes 1, and one chromosome 1 showing a deletion del (1q21-qter). In addition, three small translocations of chromosome 1 were detected. (B) Interphase nucleus of the same cell line showing three main and three small chromosome 1 domains (arrows). (Courtesy of T. Cremer, P. Emmerich, and P. Lichter, Heidelberg.)

numbers can easily be made as a result of inefficient protein removal. It is crucial therefore, to optimize pretreatment and NISH conditions for individual tissues, cells, or tumours. In cases where sections of the tumour material have to be used, it is important to realize that fragmentation of nuclei will often occur, resulting in a broad range of detectable spot numbers. We expect that in future NISH analysis on isolated intact nuclei from paraffin sections will enable the detection of the exact copy number. Detection of aneuploidy is, however, readily feasible both in tumour cell suspensions and sections; determination of the exact chromosome copy number in any given cell is often hampered by the technical difficulties mentioned above. Moreover, the evaluation of NISH data is hampered by the nature of the hybridization signals: e.g. blurred spots and split spots, can result in false aneuploidy assignment. From recent results (Masumoto *et al.* 1989) it may be suggested that cells showing split NISH spots occur in the G_2-phase of the cell cycle.

The NISH techniques that enable the efficient detection of an entire individual chromosome will provide additional information about structural aberrations, especially when these competition hybridizations are combined with centromeric probes or cosmids, phage or YAC (yeast artificial chromosomes) clones, for detection of large single-copy genes in multi-target ISH procedures (see also Chapter 10, for unique sequence detection in interphase nuclei).

Acknowledgements

We thank P. Poddighe, O. Moesker, E. van Hooren, H. Beck, and Dr W. Smeets for technical assistance and discussion. We thank Dr T. Cremer for kindly providing illustrations for the competition *in situ* suppression hybridization.

References

Angerer, R. C., Cox, K. H., and Angerer, L. M. (1985). *In situ* hybridization to cellular RNAs. *Gen. Eng.*, **7**, 43–65.

Bauman, J. G. J., Van der Ploeg, M., and Van Duijn. P. (1984). Fluorescent hybrido-cytochemical procedures: DNA.RNA hybridization *in situ*. In *Investigative microtechniques in medicine and microbiology*, (eds J. Chayen and G. Bitensky), pp. 41–48. Marcel Dekker, New York.

Brigati, D. J., Myerson, D., Leary, J. J., Spalholz, B., Travis, S., Fong, C. K. Y., Hsiung, G. D., and Ward, D. C. (1982). Detection of viral genomes in cultured cells and paraffin-embedded tissues sections using biotin-labeled hybridization probes. *Virology*, **126**, 32–50.

Burns, J., Chan, V. T. W., Jonasson, J. H., Fleming, K. A., Taylor, S., and McGee, J. O'D. (1985). Sensitive system for visualizing biotinylated DNA probes hybridized *in situ*; rapid sex determination of intact cells. *J. Clin. Pathol.*, **38**, 1085–92.

Burns, J., Redfern, D. R. M., Esiri, M. M., and McGee, J. O'D. (1986). Human and viral gene detection in routine paraffin embedded tissue by *in situ* hybridization with biotinylated probes: viral localisation in herpes encephalitis. *J. Clin. Pathol.*, **39**, 1066–73.

Burns, J., Graham, A. K., Franks, C., Fleming, K. A., Evans, M. F., and McGee, J. O'D. (1987). Detection of low copy human papilloma virus DNA and mRNA in routine paraffin sections of cervix by non-isotopic *in situ* hybridisation. *J. Clin. Pathol.*, **410**, 858–64.

Buroker, N., Bestwick, R., Haight, G., Magenis, R. E., and Litt, M. (1987). A hyper-variable repeated sequence on human chromosome 1p36. *Human Genet.*, **77**, 175–81.

Coghlan, J. P., Aldred, P., Haralambridis, J., Niall, H. D., Penschow, J. D., and Tregear, G. W. (1985). Hybridization histochemistry. *Analyt. Biochem.*, **149**, 1–28.

Cooke, H. J. and Hindley, J. (1979). Cloning of human satellite III DNA: different components are on different chromosomes. *Nucl. Acid Res.*, **6**, 3177–97.

Cooke, H. J., Schmidtke, J., and Gosden, J. R. (1982). Characterization of a repeated sequence in higher primates. *Chromosoma*, **87**, 491–502.

Cremer, T., Landegent, J., Bruckner, A., Scholl, H. P., Schardin, M., Hager, H. D., Devilee, P., Pearson, P., Van der Ploeg, M. (1986). Detection of chromosome aberrations in the human interphase nucleus by visualization of specific target DNAs with radioactive and non-radioactive *in situ* hybridization techniques: diagnosis of trisomy 18 with probe Ll.84. *Human Genet.*, **74**, 346–52.

Cremer, T., Tessin, D., Hopman, A. H. N., and Manuelidis, L. (1988a). Rapid inter-phase and metaphase assessment of specific chromosomal changes in neuroectodermal tumour cells by *in situ* hybridization with chemically modified DNA probes. *Exp. Cell Res.*, **176**, 199–206.

Cremer, T., Lichter, P., Borden, J., Ward, D. C., and Manuelidis, L. (1988*b*). Detection of chromosome aberrations in metaphase and interphase tumor cells by *in situ* hybridization using chromosome-specific library probes. *Human Genet.*, **80**, 235–46.

Dekken, H., van Pinkel, D., Mullikan, J., and Gray, J. (1988). Enzymatic production of single-stranded DNA as a target for fluorescence *in situ* hybridization. *Chromosoma*, **97**, 1–5.

Dekken, H. van and Bauman, J. G. J. (1989). A new application of *in situ* hybridization: detection of numerical and structural chromosome aberrations with a combination centromeric-telomeric DNA probe. *Cytogenet. Cell. Genet.*, **78**, 251–9.

Devilee, P., Cremer, T., Slagboom, P., Bakker, E., Scholl, H. P., Hager, H. D., Stevenson, A. F. G., Cornelisse, C. J., and Pearson, P. L. (1986). Two subsets of human alphoid repetitive DNA show distinct preferential localization in the pericentromeric heterochromatin of chromsome 13, 21, 18. *Cytogenet. Cell. Genet.*, **41**, 193–201.

Devilee, P., Kievits, T., Waye, J. S., Pearson, P. L. and Williard, H. F. (1988). Chromosome-specific alpha satellite DNA: Isolation and mapping of a polymorphic alphoid repeat from human chromosome 10. *Genomics*, **3**, 1–7.

Devilee, P., Thierry, R. F., Kievits, T., Kolluri, R., Hopman, A. H. N., Willard, H. F., Pearson, P. L., and Cornelisse, C. J. (1988). Detection of chromosome aneuploidy in interphase nuclei from human primary breast tumors using chromosome-specific repetitive DNA probes. *Cancer Res.*, **48**, 5825–30.

Dolon, T., Wyman, A. R., Mulholland, J., Barker, D., Burns, G., Latt, S., and Botstein, D. (1986). Alpha satellite-like sequences at the centromere of chromosome 8. *Am. J. Hum. Genet. Suppl.*, **39**, A196.

Emmerich, P., Jauch, A., Hofmann, M.-C., Cremer, T., and Walt, H. (1989*a*). Interphase cytogenetics in paraffin embedded sections from testicular germ cell tumor xenografts and in corresponding cultured cells. *Lab. Invest.*, (submitted).

Emmerich, P., Loos, P., Jauch, A., Hopman, A. H. N., Wiegant, J., Higgins, M., White, B. N., van der Ploeg, M., Cremer, C., and Cremer, T. (1989*b*). Double *in situ* hybridization in combination with digital image analysis: A new approach to study interphase chromosome topography. *Exp. Cell. Res.*, **181**, 126–40.

Feitz, W. F. J., Beck, H. L. M., Smeets, A. W. G. B., Debruyne, F. M. J., Vooijs, G. P., Herman, C. J., and Ramaekers, F. C. S. (1985). Tissue specific markers in flow cytometry of urological cancers: cytokeratins in bladder carcinoma. *Int. J. Cancer.*, **36**, 349–56.

Forster, A. C., McInnes, J. L., Skingle, D. C., and Symons, R. H. (1985). Non-radioactive hybridization probes prepared by the chemical labelling of DNA and RNA with a novel reagent, photobiotin. *Nucl. Acids Res.*, **13**, 745–61.

Frommer, M., Paul, C., and Vincent, P. C. (1988). Localisation of satellite DNA sequences on human metaphase chromosomes using bromodeoxyuridine-labelled probes. *Chromosoma*, **97**, 11–18.

Gerhard, D. S., Kawasaki, E. S., Bancroft, F. C., and Szabo, P. (1981). Localization of a unique gene by direct hybridization *in situ*. *Proc. Natl. Acad. Sci. USA*, **78**, 3755–9.

Giwercman, A., Hopman, A. H., N., Ramaekers, F. C. S., and Skakkebaek, N. E. (1989). Carcinoma *in situ* of the testis: detection of malignant germ cells in seminal fluid by means of *in situ* hybridization. (In preparation.)

Haase, A. T., Walker, D., Stowring, L., Ventura, P., Geballe, A., Blum, H., Brahic, M., Goldberg, R., and O'Brien, K. (1985). Detection of two viral genomes in single cell by double-label hybridization *in situ* and color microautoradiography. *Science*, **227**, 189–91.

Harper, M. E., Ullrich, A., and Saunders, G. F. (1981). Localization of human insulin gene to the distal end of the short arm of chromosome 11. *Chromosoma*, **83**, 431–9.

Harper, M. E., Marselle, L. M., Gallo, R. C., and Wong-Stahl, F. (1986). Detection of lymphocytes expressing human T-lymphotropic virus type III in lymph nodes and peripheral blood from infected individuals by *in situ* hybridization. *Proc. Natl Acad. Sci. USA*, **83**, 772–6.

Herrington, C. S. and McGee, J. O'D. (1990). Interphase cytogenetics. *Neurochem. Res.* (In press.)

Herrington, C. S., Burns, J., Graham, A. K., Evans, M., and McGee, J. O'D. (1989). Interphase cytogenetics using biotin and digoxigenin-labelled probes I: relative sensitivity of both reporter molecules for HPV 16 detection in CaSki cells. *J. Clin. Pathol.*, **42**, 592–600.

Herrington, C. S., Burns, J., Bhatt, B., Graham, A. K., and McGee, J. O'D. (1989*b*). Interphase cytogenetics using biotin and digoxigenin-labelled probes II: simultaneous differential detection of human and HPV nucleic acids in individual nuclei. *J. Clin. Pathol.*, **42**, 601–6.

Higgins, M. J., Wang, H., Shtromas, I., Haliotis, T., Roder, J. C., Holden, J. J. A., and White, B. N. (1985). Organization of a repetitive human 1.8 kb Kpnl sequence localized in the heterochromatin of chromosome 15. *Chromosoma*, **93**, 77–86.

Hopman, A. H. N., Wiegant, J., Tesser, G. I. and Van Duijn, P. (1986*a*). A non-radioactive *in situ* hybridization method based on mercurated nucleic acid probes and sulfhydryl-hapten ligands. *Nucl. Acids Res.*, **14**, 6471–88.

Hopman, A. H. N., Wiegant, J., and Van Duijn. P. (1986*b*). Mercurated nucleic acid probes, a new principle for non-radioactive *in situ* hybridization. *Exp. Cell. Res.*, **169**, 357–68.

Hopman, A. H. N., Wiegant, J., Raap, A. K., Landegent, J. E., Van der Ploeg, M., and Van Duijn, P. (1986*c*). Bi-color detection of two target DNAs by non-radioactive *in situ* hybridization. *Histochemistry*, **85**, 1–4.

Hopman, A. H. N., Wiegant, J., and Van Duijn, P. (1986*d*). A new hybridocyto-chemical method based on mercurated nucleic acid probes and sulfhydryl-hapten ligands. I. Stability of the mercury-sulfhydryl bond and influence of the ligand structure on the immunochemical detection of the hapten. *Histochemistry*, **84**, 169–78.

Hopman, A. H. N., Ramaekers, F. C. S., Raap, A. K., Beck, J. L. M., Devilee, P., Van der Ploeg, M., and Vooijs, G. P. (1988*b*). *In situ* hybridization as a tool to study numerical chromosome aberrations in solid bladder tumors. *Histochemistry*, **89**, 307–16.

Hopman, A. H. N., Raap, A. K., Landegent, J. E., Wiegant, J., Boerman, R. H., and van der Ploeg, M. (1988*d*). Non-radioactive *in situ* hybridization. In *Molecular neuroanatomy*, (eds F. W. Van Leeuwen, R. M. Buijs, C. W. Pool, and O. Pach), pp. 43–68. Elsevier, Amsterdam.

Hopman, A. H. N., Poddighe, P., Smeets, W., Moesker, O., Beck, J., Vooijs, G. P., and Ramaekers, F. C. S. (1989). Detection of chromosome #1 and #18 aberrations in bladder cancer by *in situ* hybridization. A comparison with flow cytometric data. *Am. J. Pathol.*, (in press).

Kievits, T., Devilee, P., Wiegant, J., Wapenaar, M. C., Cornelisse, C., van Ommen, G. J. B., and Pearson, P. L. (1989). Direct non-radioactive *in situ* hybridization of somatic cell hybrids DNA to human lymphocytes chromosomes. *Cytometry*, (in press).

Landegent, J. E., Jansen in de Wal, N., Baan, R. A., Hoeijmakers, J. H. J., and Van der Ploeg, M. (1984). 2-acetylaminofluorene-modified probes for the indirect hybridocytochemical detection of specific nucleic acid sequences. *Exp. Cell. Res.*, **153**, 61–72.

Landegent, J. E., Jansen in de Wal, N., Ploem, J. S., and Van der Ploeg, M., (1985). Sensitive detection of hybridocytochemical results by means of reflection-contrast microscopy. *J. Histochem. Cytochem.*, **33**, 1241–6.

Landegent, J. E., Jansen in de Wal, N., Dirks, R. W., Baas, F., van der Ploeg, M. (1987). Use of whole cosmid cloned genomic sequences for chromosomal localization by non-radioactive *in situ* hybridization. *Human Genetics*, **77**, 366–70.

Langer, P. R., Waldrop, A. A., and Ward, D. C. (1982). Enzymatic synthesis of biotin labeled polynucleotides: novel nucleic acid affinity probes. *Proc. Natl Acad. Sci. USA*, **78**, 6633–7.

Lawrence, J. B., Singer, R. H., and Marselle, L. M. (1989). Highly localized tracks of specific transcripts within interphase nuclei visualized by *in situ* hybridization. *Cell*, 493–502.

Lichter, P., Cremer, T., Borden, J., Manuelidis, L., and Ward, D. C. (1988). Delineation of individual human chromosomes in metaphase and interphase cells by *in situ* suppression hybridization using recombinant DNA libraries. *Human Genet.*, **80**, 223–4.

Lichter, P., Cremer, T., Chang Tang, C. J., Watkins, P. C., Manuelidis, L., and Ward, D. C. (1989). Rapid detection of human chromosome 21 aberrations by *in situ* hybridization. *Proc. Natl. Acad. Sci. USA*, (in press).

McGee, J. O'D., Burns, J., and Fleming, K. A., (1987). New molecular techniques in pathological diagnosis: visualisation of genes and mRNA in human tissue. In *Investigational techniques in oncology*, (ed. N. M. Bleehen), pp. 21–34. Springer-Verlag, London.

Manuelidis, L. (1985). Individual interphase chromosome domains revealed by *in situ* hybridization. *Human Genet.*, **71**, 288–93.

Manuelidis, L. and Ward, D. C. (1985). Chromosomal and nuclear distribution of the Hind III, 1.9-kb human repeat segment. *Chromosoma*, **91**, 28–38.

Masumoto, H., Sugimoto, K., and Okazaki, T. (1989). Alphoid satellite DNA is tightly associated with centromere antigens in human chromosomes throughout the cell cycle. *Exp. Cell. Res.*, **181**, 181–96.

Matthews, J. A. and Kricka, L. J. (1988). Analytical strategies for the use of DNA probes. *Analyt. Biochem.*, **169**, 1–25.

Moyzis, R. K., Albright, K. L., Bartholdi, M. F., Cram, L. S., Deaven, L. L., Hildebrand, C. E., Joste, N. E., Longmire, J. L., Meyne, J. and Schwarzacher-

Robinson, T. (1987). Human chromosome-specific repetitive DNA sequences: Novel markers for genetic analysis. *Chromosoma*, **95**, 375–82.

Naoumov, N. V., Alexander, G. J. M., Eddleston, A. L. W. F. and Williams, R. (1988). *In situ* hybridisation in formalin fixed, paraffin wax embedded liver specimens: method for detecting human and viral DNA using biotinylated probes. *J. Clin. Pathol.*, **41**, 793–8.

Nederlof, P. M., Robinson, D., Abuknesha, R., Wiegant, J., Hopman, A. H. N., Tanke, H. J. and Raap, A. K. (1989*a*). Three-color-fluorescence *in situ* hybridization for the simultaneous detection of multiple nucleic acid sequences. *Cytometry*, **10**, 20–8.

Nederlof, P. M., Van der Flier, S., Raap, A. K., Tanke, H. J., Van der Ploeg, M., Kornips, F., and Geraedts, J. P. M. (1989*b*). Detection of chromosome aberrations in interphase tumor nuclei by non-radioactive *in situ* hybridization. *Cancer Genet. Cytogenet.*, (in press).

Nederlof, P., van der Flier, S., Wiegant, J., Raap, A. K., Tanke, H. I., Ploem, J. S., and van der Ploeg, M. (1989*c*). Multiple-fluorescence *in situ* hybridization. *Cytometry*, (in press).

Pezzella, M., Pezzella, F., Galli, C., Macchi, B., Verani, P., Sorice, F., and Baroni, C. D. (1987). *In situ* hybridization of human immunodeficiency virus (HTLV-III) in cryostat sections of lymph nodes of lymphadenopathy syndrome patients. *J. Med. Virol.*, **22**, 135–42.

Pinkel, D., Straume, T. and Gray, J. (1986). Cytogenetic analysis using quantitative, high-sensitivity, fluorescence hybridization. *Proc. Natl Acad. Sci. USA*, **83**, 2934–8.

Pinkel, D., Landegent, J., Collins, C., Fuscoe, J., Segraves, R., Lucas, J., and Gray, J. (1988). Fluorescence *in situ* hybridization with human chromosome-specific libraries: Detection of trisomy 21 and translocations of chromosome 4. *Proc. Natl Acad. Sci. USA*, **85**, 9138–42.

Raap, A. K., Marijnen, J. G. J., Vrolijk, J., and Van der Ploeg, M. (1986). Denaturation, renaturation, and loss of DNA during *in situ* hybridization procedures. *Cytometry*, **7**, 235–42.

Raap, A. K., Geelen, J. L., Van der Meer, J. W. M., Van de Rijke, F. M., Van den Boogaard, P., and Van der Ploeg, M. (1988). Non-radioactive *in situ* hybridization for the detection of cytomegalovirus infections. *Histochemistry*, **88**, 367–73.

Raap, A. K., Hopman, A. H. N., and Van der Ploeg, M. (1989). Hapten labeling of nucleic acid probe for DNA *in situ* hybridization. In *Techniques in immunocytochemistry*, (ed. G. R. Bullock) Vol. 4, (in press).

Rappold, G., Cremer, T., Hager, H., Davies, K., Muller, C., and Yang, T. (1984). Sex chromosome positions in human interphase nuclei as studied by *in situ* hybridization with chromosome specific DNA probes. *Human Genet.*, **67**, 317–25.

Schardin, M., Cremer, T., Hager, H. D., and Lang, M. (1985). Specific staining of human chromosomes in Chinese hamster × man hybrid cell lines demonstrates interphase chromosome territories. *Human Genet.*, **71**, 281–7.

Singer, R. H., Lawrence, J. B., Langevin, G. L., Rashtchian, R. N., Villnave, C. A., Cremer, T., Tesin, D., Manuelidis, L., and Ward, D. C. (1987). Double labelling *in situ* hybridization using non-isotopic and isotopic detection. *Acta Histochem.*

Cytochem., **20**, 589–99.

Singer, M. (1982). Highly repeated sequences in mammalian genomes. *Int. Rev. Cytol.*, **76**, 67–112.

Smeets, A. W. G. B., Pauwels, R. P. E., Beck, H. L. M., Feitz, W. F. J., Geraedts, J. P. M., Debruyne, F. M. J., Laarakkers, L., Vooijs, G. P., and Ramaekers, F. C. S. (1987). Comparison of tissue disaggregation techniques of transitional cell bladder carcinomas for flow cytometry and chromosomal analysis. *Cytometry*, **8**, 14–19.

Tchen, P., Fuchs, R. P. P., Sage, E., and Leng, M. (1984). Chemically modified nucleic acids as immunodetectable probes in hybridization experiments. *Proc. Natl Acad. Sci. USA*, **81**, 3466–70.

Tourtelotte, W., Verity, A. N., Schmid, P., Martinez, S., and Shapshak, P. (1986). Covalent binding of formalin fixed paraffin embedded brain tissue sections to glass slides suitable for *in situ* hybridization. *J. Virol. Methods*, **15**, 87–99.

Trask, B., Engh, G. van den, Pinkel, D., Mullikin, J., Waldman, F., Dekken, H. van, and Gray, J. (1988). Fluorescence *in situ* hybridization to interphase cell nuclei in suspension allows flow cytometric analysis of chromosome content and microscopic analysis of nuclear organization. *Human Genet.*, **78**, 251–9.

van der Ploeg, M. and Raap, A. K. (1988). *In situ* hybridization: An overview. In *New frontiers in cytology*, (ed. K. Goerttler, G. E. Feichter, and S. Witte), pp. 13–21. Springer-Verlag, New York.

Van Prooijen-Knegt, A. C., Van Hoek, J. F. M., Bauman, J. G. J., Van Duijn, P., Wool, I. G., and Van der Ploeg, M. (1982). *In situ* hybridization of DNA sequences in human metaphase chromosomes visualized by an indirect fluorescent immuno-cytochemical procedure. *Exp. Cell Res.*, **141**, 397–407.

Waye, J. S. and Willard, H. F. (1986). Molecular analysis of a deletion polymorphism in alpha satellite of human chromosome 17: evidence for homologous unequal crossing-over and subsequent fixation. *Nucl. Acids Res.*, **14**, 6915–28.

Waye, J. S., Durfy, S. J., Pinkel, D., Kenwrick, S., Patterson, M., Davies, K., and Willard, H. F. (1987*a*). Chromosome specific alpha satellite DNA from human chromosome 1: Hierarchical structure and genomic organization of a poly-morphic domain spanning several hundred kilobase pairs of centromeric DNA. *Genomics*, **1**, 43–52.

Waye, J. S., England, S. B., and Willard, H. F. (1987*b*). Genomic organization of alpha satellite DNA on human chromosome 7: Evidence for two distinct alphoid domains on a single chromosome. *Mol. Cell. Biol.*, **7**, 349–56.

Willard, H. F., Smith, K. D., and Sutherland, J. (1983). Isolation and characterization of a major tandem repeat family from the human X chromosome. *Nucl. Acids Res.*, **11**, 2017–33.

Willard, H. F. and Waye, J. S. (1987). Hierarchical order in chromosome-specific human alpha satellite DNA. *Trends in Genetics*, **3**, 192–8.

12

Single and simultaneous nucleic acid detection in archival human biopsies

APPLICATION OF NON-ISOTOPIC *IN SITU* HYBRIDIZATION AND THE POLYMERASE CHAIN REACTION TO THE ANALYSIS OF HUMAN AND VIRAL GENES

C. S. HERRINGTON, D. M. J. FLANNERY,
AND J. O'D. McGEE

Introduction

In situ hybridization (ISH) may be defined as the direct detection of nucleic acid in intact cellular material. Its application to human pathology therefore involves the detection of both normal and abnormal nucleic acids in human cells and tissues (for review see Warford 1988). In the context of pathology, development of techniques for ISH has been directed towards procedures that are clinically useful. This requires not only the production of clinically relevant information but also the ability to perform the procedures as part of a routine diagnostic service. A variety of cell and tissue samples can be studied using ISH, from individual chromosomes in metaphase spreads (Bhatt *et al.* 1988) to archival paraffin-embedded biopsy material. Using appropriately labelled probes, the presence or absence of normal and abnormal nucleic acids can be detected and correlated with cell and tissue morphology. For many years, ISH was performed using isotopic probe labels for the detection of hybridized duplexes. However, these require autoradiographic detection techniques, which are time-consuming. More recently, isotopes of higher specific activity, which require shorter autoradiographic exposure times, have been used, e.g. [35]S (see Chapter 3). A quicker and safer approach is to use non-isotopic probe labels, detected by histochemical means. This is termed non-isotopic *in situ* hybridization (NISH). Many different compounds have been used as probe labels (Matthews and Kricka 1988) but the most widely used is biotin (Langer *et al* 1981). This was originally employed in

NISH in order to utilize its high-affinity binding to the naturally occurring protein avidin or streptavidin. Many different methods of detection have been developed, utilizing conjugates of avidin to both enzymes and fluorochromes. More recently, alternatives to biotin have been explored in order to circumvent the problem of endogeneous biotin that is present in many tissues. In parallel, techniques have been developed for detection of more than one nucleic acid within individual cells. This has been achieved in isolated cells (Hopman *et al.* 1988; Cremer *et al.* 1988; Nederlof *et al.* 1989) using fluorescent detection systems (see Chapter 11) and more recently in archival biopsy material using non-fluorescent detection (Herrington *et al.* 1989*a,b*). This allows analysis of the relationship between nucleic acids within cells and the morphology of the tissue containing them. It is important to note that NISH on paraffin sections requires non-fluorescent detection of probe labels because fluorescent methods are less suitable owing to tissue autofluorescence. In addition, fluorescent detection systems are less applicable to routine testing; many fluorescent molecules fade on exposure to light.

NISH has been applied to the analysis of both human and non-human DNA in clinical biopsies (Beckmann *et al.* 1985; Burns *et al.* 1986). However, the flexibility and clinical applicability of the techniques are best illustrated with reference to viral detection. The remainder of this chapter is therefore concerned with viral detection in clinical samples.

Viral detection in clinical samples

Introduction

The diagnosis of viral infection can often be inferred from clinical features. However, definite diagnosis requires either direct demonstration of the virus or its cytopathic effect, or detection of a specific immunological response. The latter, however, depends on the phase of infection, immunocompetence of the host, and assays for humoral and/or cell-mediated immunity. In addition, widespread exposure of the population to certain types of virus, e.g. Epstein–Barr virus (EBV), can complicate diagnosis, as only a specific IgM response is informative. Similarly, anamnestic rises in antibody titre can create diagnostic confusion. The direct demonstration of virus by viral culture or within tissues or body fluids provides a definite diagnosis. For some viruses, viral culture is either impossible (e.g. for human papillomaviruses, HPV) or technically time-consuming. Viruses can be demonstrated directly in human tissues in several ways. These include conventional histological staining, electron microscopy, and immunohistochemistry for viral antigens. The use of morphological criteria, that is, the detection of the cytopathic effect of the virus, for the diagnosis of viral infection is non-specific and insensitive. For example, koilocytes are not indicative of infection by a particular type of HPV and, similarly, not all lesions containing HPVs show koilocytic atypia. Immunohistochemistry detects only the antigen to which the antibody is directed and does not direct virus that is not producing that antigen. CaSki cells, for example, which contain integrated HPV16 sequences, do not express viral capsid protein and therefore do not stain with antibodies to that protein. Electron microscopy is useful for the diagnosis of specific lesions, e.g. atypical herpetic vesicular eruptions, but is only capable of detecting intact virions; additionally, viral subtypes, e.g. HPV or herpes viruses, look the same and are therefore indistinguishable.

Viruses are defined, classified, and typed according to their nucleic acid content (Taussig 1984). The detection of nucleic acid in clinical material is therefore the most appropriate way of achieving a clinical diagnosis and of investigating the epidemiology and natural history of viral infections. Nucleic acids can be analysed in two ways: after extraction from tissue, with or without amplification by the polymerase chain reaction (PCR), or directly by *in situ* hybridization. In the remainder of this chapter, only *in situ* hybridization, and PCR detection of HPV as it relates to NISH sensitivity, will be discussed.

Viruses may contain DNA or RNA but not both. DNA may be double- or single-stranded, and RNA in the sense or antisense form. Finally, an RNA virus may replicate via a double-stranded DNA intermediate. These considerations form the basis of the Baltimore classification (Taussig 1984) and are a logical way in which to consider viral detection by ISH. Many viruses have now been investigated using ISH (Grody *et al.* 1987; McDougall 1988). Those receiving the most attention have been the DNA viruses: HPV (Syrjanen *et al.* 1988), herpesviruses (EBV, HSV, CMV) (Burns *et al.* 1986; Wolber *et al.* 1988), and polyomaviruses (JC, BK) (Boerman *et al.* 1989), which are double-stranded; parvoviruses (Porter *et al.* 1988), which are single-stranded; and hepatitis B virus (Blum *et al.* 1983), which has a mixed double- and single-stranded genome. Most DNA viruses replicate within the cell nucleus (the notable exception being poxviruses) and the nucleic acid may remain separate from the host genome, i.e. episomal, or may integrate into one or more chromosomes. This difference may be detectable by *in situ* hybridization by differences in signal morphology. Episomal viruses are replicating and are therefore present in large numbers. This facilitates their detection by *in situ* hybridization. However, integrated viruses may be present in very low numbers and therefore require sensitive procedures for their detection. The sensitivity of non-isotopic procedures presently in routine use is approximately 30 copies (Burns *et al.* 1987; Syrjanen *et al.* 1988; Walboomers *et al.* 1988; Herrington *et al.* 1989c) and therefore the integration of virus in numbers lower than this may not be detectable. However, the amplification of detection procedures for the detection of non-isotopic labels in archival material increases the sensitivity up to tenfold (Herrington and McGee, unpublished) and the detection of low-copy-number viruses is therefore possible but not yet routinely applicable. Another approach to the problem of low copy number is the detection of virus-specific mRNA, which is not only specific for the virus but also for the phase of infection. However, this provides no information regarding the physical state of the viral genome.

RNA viruses are also detectable by ISH (Haase *et al.* 1985*a*,*b*, 1986) but these have received less attention until recently. The intact virion of oncornaviruses contains single-stranded RNA but a double-stranded DNA intermediate is formed during replication. This intermediate integrates into the host genome and has been implicated in oncogenesis. HIV-1 belongs to this group of viruses and has been investigated by NISH (Pezzella *et al.* 1987).

HPVs have been extensively investigated by all methods described above and illustrate the ways in which NISH can be used in this context. Therefore, the detection of HPV in biopsies and cervical smears will be used as illustrative examples.

HPV detection in archival biopsies

Papillomaviruses have been detected in most lesions of squamous epithelia (Syrjanen 1987). They are known to integrate into the host genome (Popescu *et al.* 1987*a*,*b*; Mincheva *et al.* 1987) and have been implicated in the aetiology of many squamous-cell carcinomata, particularly that of the uterine cervix (Bonfiglio 1988). The virus is easily detectable by *in situ* hybridization in benign, premalignant, and malignant lesions of many sites and it has been noted that the morphological characteristics of episomal virus disappear with progression of morphology from benign to malignant. This has been taken to imply integration of the viral genome.

In situ hybridization allows the correlation of the presence of virus with the morphology of the infected tissues. By using specific probes, viral subtypes can be distinguished and the precise infecting species found. This has considerable epidemilogical importance as well as relevance for the natural history of HPV infections. Thus, it has been shown that HPV1 is associated with benign cutaneous lesions and several other less-common HPV types with the rare premalignant syndrome epidermodysplasia verruciformis (Pfister and Fuchs 1987). Similarly, HPV 6 and 11 are found in benign anogenital lesions and HPV 16, 18, 31, 33, and 35 in premalignant lesions of both male and female genitalia (Pfister and Fuchs 1987). More recently, HPV nucleic acid has been found by NISH in other squamous-cell carcinomata (SCC), such as SCC of the bronchus (Syrjanen *et al.* 1989).

The investigation of archival biopsies requires accessibility of target nucleic acid for hybridization with molecular probes. This requires initial dewaxing of sections by conventional means followed by proteolysis to remove both viral and cellular proteins that impede diffusion of probe and access of probe to target. This process is termed unmasking and has been performed with a variety of enzymes. These enzymes vary in potency, not only from type to type but also from manufacturer to manufacturer. Thus, conditions must be optimized using enzymes from one supplier. Initially pepsin-HCl was thought to be the most appropriate enzyme for unmasking HPV (Burns *et al.* 1987), primarily because the use of more potent enzymes led to an unacceptable rate of detachment of sections from slides. The use of aminopropyltriethoxysilane as a section adhesive (Burns *et al.* 1988) has allowed the use of higher concentrations of proteinase K (up to 1 mg/ml), producing significantly more effective unmasking. This is now the method used by several groups for HPV sequences (Wells *et al.* 1987; Burns *et al.* 1988; Syrjanen *et al.* 1988). However, this does not imply general utility for nucleic acid unmasking: the detection of HPV mRNA (Burns *et al.* 1987) and snail mRNA (Dirks *et al.* 1989) is optimum with pepsin-HCl. Thus, unmasking procedures must be optimized for each experimental system.

Having exposed the sequence of interest, appropriately labelled probe(s) (see Chapters 3, 4) is added and the tissue section and probe are simultaneously denatured by heat. The addition of exogenous unlabelled human or salmon sperm DNA in excess is often necessary to reduce non-specific background but has not been found beneficial for the detection of HPV in archival biopsies (see Appendix). Hybridization is then allowed to occur under appropriate conditions (for full discussion see Chapter 1 and Herrington *et al.* 1989*b*). Following hybridization, excess probe is removed and mismatched hybrids are dissociated using appropriate salt, formamide, and temperature combinations to a predetermined level of stringency (see Chapter 1). After a blocking step, probe–target hybrids are detected by non-fluorescent histochemical means.

The detection systems employed depend on the reporter molecule to be detected. Most reporters (e.g. digoxigenin, acetaminofluorene, mercury) are detected using specific antibodies, but biotin can be visualized using either antibody- or avidin-based systems. Single-step detection requires the linkage of antibody or avidin to an indicator enzyme: the most widely used are horseradish peroxidase and alkaline phosphatase. Amplified detection systems employ a variety of mechanisms to enhance signal and hence sensitivity and will be discussed below. Several substrates of different colours can be used for each of these enzymes and can be manipulated to allow either appropriate counterstaining or combination for multiple nucleic acid labelling (see below).

The sensitivity of a particular non-fluorescent detection system depends on several parameters: the availability of target sequences for hybridization, i.e. unmasking; the type of probe; the reporter molecule used; the affinity of the antibody/avidin for the reporter; the degree of amplification by multiple antibody/avidin/enzyme layers and the enzyme/chromogen combination used. The investigation of the contribution of each of these factors to sensitivity requires that all other parameters are comparable (Chapter 11).

The probes used for NISH are generally of three main types: nick-translated, random-primed, and oligonucleotide. Nick-translated probes are the most widely used as they are simple to produce and use. As the labelling method (see Chapter 4) introduces nicks at random into the double-stranded DNA molecule, labelled fragments of different length and containing variable proportions of insert and vector are produced. It has been suggested that these fragments anneal together via their overlapping portions to produce a probe network, thereby amplifying the target sequence *in situ.* The same arguments apply to probes generated by random primed synthesis but not to oligonucleotide probes. In addition, individual oligonucleotides, being shorter than other probes, are likely to be less sensitive than their nick-translated and random-primed counterparts. However, oligonucleotides are naturally more specific and can be used individually or in combination (see Chapter 5). The type of probe should therefore be chosen to suit the experimental system under study.

The affinity of the detector molecule for the reporter also contributes to sensitivity: the high affinity of avidin for biotin was one reason why this molecule was chosen as a non-isotopic probe label (Langer 1981). The maximum affinity of antibodies for antigen ($K_d = 10^{-12}$) is approximately 1000 times less than that of avidin for biotin ($K_d = 10^{-15}$). The sensitivity of single-step detection of digoxigenin-labelled probes was found to be less than that of a corresponding system for the detection of biotin (Herrington *et al.* 1989*c*). However, digoxigenin was detected using an antibody con-jugate, whereas biotin was detected using the avidin conjugate as shown in Fig. 12.1(a). The use of three-step amplification systems (Fig. 12.1(b)) employing monoclonal antibodies for both digoxigenin and biotin and the same second and third steps (see Appendix) renders the two reporters equally sensitive (Herrington and McGee, unpublished).

Amplification of detection can be achieved by either antibody layering techniques (Fig. 12.1(b)) or by enzyme–antibody complexes (Fig. 12.1(c)) that increase the number of enzyme molecules bound to each reporter molecule. Thus, the combination of monoclonal antibody to the reporter molecule (e.g. monoclonal anti-biotin used in the Appendix) linked via a biotinylated linker antibody to an avidin conjugate produces high speci-ficity (monoclonal antibody) and sensitivity (avidin conjugate). Alter-natively, monoclonal antibodies to reporter molecules can be linked through rabbit anti-mouse immunoglobulin to a complex of detector enzyme and antibody to the detector enzyme (e.g. peroxidase anti-peroxidase (PAP) and alkaline phosphatase anti-alkaline phosphatase (APAAP)) (Sternberger *et al.* 1970; Cordell *et al.* 1984). The two approaches of antibody and enzyme enhancement are combined in methods based on the avidin–biotinylated alkaline phosphatase complex

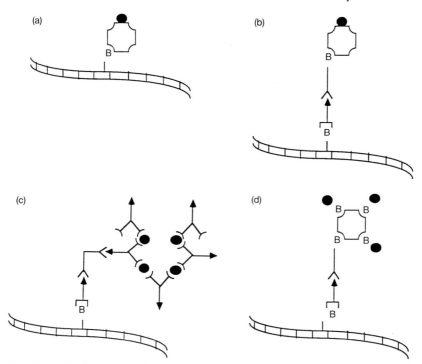

Fig. 12.1. (a) Single-step detection of biotin (B). Biotin is detected using a conjugate of avidin (○) and alkaline phosphatase (●). (b) Three-step detection of biotin. A monoclonal antibody to biotin (]→) is linked via biotinylated rabbit anti-mouse immunoglobulin (B—<) to avidin alkaline phosphatase conjugate. This combines the high specificity of the monoclonal antibody with the high-sensitivity interaction between avidin and biotin. (c) Alkaline phosphatase anti-alkaline phosphatase (APAAP) detection. A monoclonal antibody to biotin is detected using rabbit anti-mouse antibody. The second valency of this molecule is then used to bind to the preformed APAAP complex. The APAAP complex is produced by mixing alkaline phosphatase with monoclonal anti alkaline phosphatase (⋎→). (d) Avidin–biotinylated alkaline phosphatase complex (ABC). In this method, biotin is detected using a combination of antibody and enzyme amplification. Antibody to biotin is detected using a biotinylated second antibody followed by a preformed complex of avidin with biotinylated alkaline phosphatase (B●). This combines the high affinity of the avidin biotin interaction with the occupancy of the four binding sites of avidin for biotin.

(ABC) initially proposed for avidin–biotinylated peroxidase by Hsu *et al.* (1981) for the detection of antigen. These methods also exploit the tetravalency of avidin. Methods 1(b)–(d), shown in Fig. 12.1, can be modified for the detection of other reporter molecules by using the appropriate primary antibody and can be manipulated to produce combined detection systems of great flexibility. Sensitivity depends also on the choice of chromogenic enzymes and substrates: alkaline phosphatase-based systems tend to be of greater sensitivity than those employing peroxidase unless enhanced substrates are used, e.g. metal-chelated DAB (Hsu *et al.* 1982).

Similarly, silver enhancement of diaminobenzidine (DAB) significantly increases the sensitivity of peroxidase-based detection of sequences in archival material (Burns *et al.* 1985).

The measurement of the sensitivity of NISH detection systems is not easy. Relative sensitivity can be assessed by estimation of the number of positive cells within a given area of comparable tissue sections using different systems. Absolute sensitivity can only be estimated by analysing the ability of particular systems to detect sequences of known copy number. This has been performed using cell lines containing multiple copies of HPVs, for example CaSki, HeLa, and SiHa cells. Syrjanen *et al.* (1988) estimated the sensitivity of biotinylated probes to be 10–50 copies on the basis of their ability to detect HPV consistently in CaSki and HeLa but not SiHa cells. However, the copy number is likely to vary from cell to cell and a more appropriate method of estimation is to analyse the frequency distribution of signals within these cells (for example, see Fig. 12.2) and compare the median of the distribution with the average copy number per cell derived by dot blot hybridization (Herrington *et al.* 1989*c*). A third approach is to compare the ability of a given system to produce a signal within tissue sections with the copy number derived by dot blot hybridization of DNA extracted from a parallel sample (Walboomers *et al.* 1988). An internal non-viral control of known copy number can also be used (Burns *et al.* 1987). It is of note that all of these methods have given estimates for the sensitivity of biotinylated probes of approximately 30 copies of HPV per cell.

The single-step detection method described here is useful for the rapid

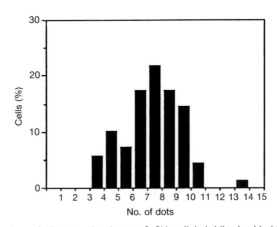

Fig. 12.2. The number of discrete signals per CaSki cell hybridized with biotinylated HPV16 probe has been plotted against the percentage of cells. The distribution of dot number in 80 cells is skewed with a median of 8 dots per cell. The copy number of mid log phase CaSki cells was estimated to be 270 by dot blot hybridization. This gives a sensitivity of approximately 30 copies.

screening of tissue blocks for the presence of HPV infection (Fig. 12.1(a) and 12.3). This indicates the presence of high-copy-number infection of cells, as the sensitivity of the procedure has been estimated to be of the order of 30 copies of virus per cell (see above). The amplified detection procedure increases the sensitivity of detection up to tenfold (Herrington and McGee, unpublished) and will therefore give additional information regarding low-copy-number infections.

It has been suggested that the analysis of material by PCR has greater diagnostic sensitivity than NISH, and several groups have analysed archival material in this way (Shibata *et al.* 1988; Herrington *et al.* 1989*b*). PCR is an *in vitro* molecular cloning technique that can amplify a specific DNA region by as much as 10^8-fold (Saiki *et al.* 1985). In essence, nucleic acids are exposed by dewaxing and boiling. Cycles of primer-directed synthesis are then performed (see Fig. 12.4) using a thermally stable DNA polymerase from *Thermus aquaticus* (Taq). Taq polymerase effectively shuttles back and forth between primers, producing an exponentially increasing number of copies of the intervening sequences. The resultant DNA is analysed both by agarose gel electrophoresis with ethidium bromide staining and by Southern transfer followed by hybridization to an internal oligonucleotide probe (Fig. 12.5).

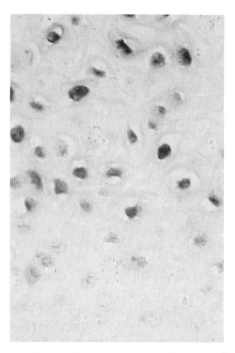

Fig. 12.3. A section from a formalin-fixed paraffin-embedded condyloma acuminatum was hybridized with biotinylated HPV6 probe. Clear nuclear signal can be seen.

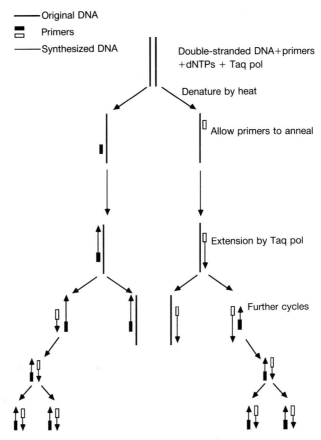

——— Original DNA
▣ Primers
——— Synthesized DNA

Double-stranded DNA+primers
+dNTPs + Taq pol

Denature by heat

Allow primers to anneal

Extension by Taq pol

Further cycles

Fig. 12.4. The principle of the polymerase chain reaction. The two primers are chosen in order to produce specificity of amplification. The reaction product is of predictable molecular size.

The ability of this approach to detect HPV sequences in archival biopsies has been compared with NISH using biotinylated probes detected with the amplified detection system (see Appendix). As shown in Table 12.1, there is complete concordance between the two methods and the diagnostic sensitivity of NISH is therefore equivalent to that of PCR. The absolute sensitivity can be estimated by counting the number of positive cells in a given biopsy and comparing this with the total number of cells in the section. This gives a sensitivity for NISH of 1 in 10 000 cells (Graham and McGee, unpublished).

NISH analysis allows the individual identification of HPV subtypes in clinical biopsies and hence the distinction of 'benign' and 'malignant' viral subtypes. The application of these methods to archival biopsy files will provide a wealth of retrospective information regarding HPV infection of human tissues. Indeed, a study is at present under way of the cases analysed

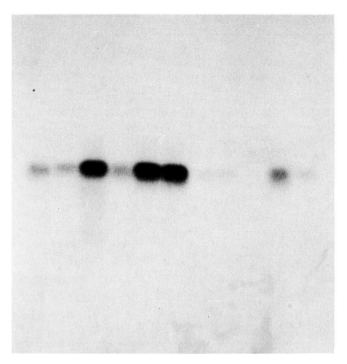

Fig. 12.5. The bands shown represent HPV16 specific sequences generated by PCR from 10 different archival biopsies. These were produced by hybridization on nitrocellulose filter generated by the Southern transfer of an agarose gel with a biotinylated internal oligonucleotide probe (see text).

epidemiologically by Sir Richard Doll. Correlation of the presence of HPV with the epidemiological parameters and subsequent clinical course of the patients will give valuable information regarding the natural history of HPV infection of the cervix and its relationship to cervical cancer.

Analysis of cervical smears

Although the analysis of cervical biopsies by NISH is useful in the investigation of HPV infection, the necessary prerequisite for this analysis is that the patient has had a cervical biopsy. However, cervical screening for premalignant lesions involves the collection of cervical smears, which are analysed cytologically for the presence of abnormal cells indicative of cervical intra-epithelial neoplasia. The detection of HPV DNA in cervical smears has posed several problems. Firstly, only one smear is taken and this is naturally required for cytopathological diagnosis. It is thus impossible to perform controls on material from the same patient. Secondly,

Table 12.1. NISH and PCR Genotyping of HPV

Case no.	Tissue	NISH	PCR*	Diagnosis
1	Wax only	0	0	—
2	Tonsil	0	0	Normal
3	CaSki cells	16	16	Cell line
4	Cervix	16	16	CIN3†
5	Cervix	16	16	CIN3
6	Cervix	16	16	CIN3
7	Cervix	16	16	CIN2
8	Cervix	16	16	ISCC
9	Cervix	16	16	CIN3
10	Cervix	16	16	ISCC‡
11	Cervix	16	16	Condyloma
12	Cervix	16	16	ISCC
13	Cervix	6	0	CIN2
14	Cervix	6	0	Normal
15	Cervix	6	0	CIN3
16	Cervix	11	0	CIN2/3
17	Cervix	11	0	CIN2
18	Cervix	11	0	CIN1/2
19	Cervix	33	0	CIN1
20	Cervix	16	16	CIN2
21	Cervix	16	16	Condyloma
22	Cervix	0	0	ISCC

* PCRs were done only for HPV 16.
† CIN, cervical intraepithelial neoplasia.
‡ ISCC, invasive squamous cell carcinoma.

cervical smears contain a variable amount of mucin and are contaminated with bacteria, fungi, and protozoa. This may lead to technical problems, particularly high background staining.

HPV infection of exfoliated cells has been analysed by a variety of methods. The analysis of smears from normal women by Southern blotting suggested that 11–12 per cent of normal smears harboured HPV infection (Lorincz *et al.* 1986; Toon *et al.* 1986). 'Filter *in situ* hybridization' has provided evidence in agreement with this (Wagner *et al.* 1984) but a recent large study of 1930 normal smears showed that only 1 per cent of these samples contained HPV sequences (Melchers *et al.* 1989). It is of note that these authors found 100 per cent concordance between filter *in situ* hybridization and Southern blot analysis. The percentage of smears containing HPV sequences increases with the degree of cytological abnormality from normality to CINIII. In addition, the proportion of cases infected with HPV16/18 increases with greater cytological abnormality. This has been noted using both Southern blot analysis (Lorincz *et al.* 1986; Burk *et al.* 1986) and filter *in situ* hybridization (Wagner *et al.* 1984; De Villiers *et al.* 1987).

A recent study by *in situ* hybridization has shown agreement with the studies quoted above: 15.4 per cent of normal smears contained HPV sequences and the percentage positive increased with increasing abnormality to a maximum of 78.2 per cent for CINIII. The sensitivity of the NISH method used was estimated as 50–100 copies and was judged to be greater than that of Southern blot analysis (Pao *et al.* 1989)! The comparison of NISH with 'Verapap', which is a modification of filter *in situ* hybridization, has shown a moderate degree of agreement between the two methods (Table 12.2). Verapap, however, gives no morphological information.

Table 12.2. Comparison of Verapap and NISH for the detection of HPV in cervical smears

		Verapap	
		+	−
NISH	+	13	11
	−	7	68

However, the application of the polymerase chain reaction to this problem has demonstrated a much higher incidence of HPV infection in normal smears, namely 70 per cent (Young *et al.* 1988). Moreover, no correlation was found between the presence of HPV and the cytological grading, as all abnormal smears were positive for HPV. Similarly, there was no higher incidence of HPV16/18 infection with the higher grades of CIN. Preliminary results of NISH on smears taken from women attending a sexually transmitted disease clinic (Fig. 12.6) has, however, shown a high infection rate of cytologically normal smears (Herrington and McGee, unpublished).

Thus, the investigation of HPV infection of exfoliated cervovaginal cells has produced conflicting results. The application of more refined methodology will be required to resolve these difficulties.

Simultaneous double nucleic acid detection in archival biopsies and cervical smears

Introduction

Techniques for the histochemical detection of two antigens (Wagner *et al.* 1988) or antigen and nucleic acid (Wolber *et al.* 1988) have been described and the simultaneous detection of two nucleic acids has been performed by isotopic *in situ* hybridization using two isotopic labels (Haase *et al.* 1985). This, however, requires the use of differential autoradiography with its attendant lack of cellular resolution. It also has the disadvantage that the operator waits weeks for a result and is exposed to a dual radiation hazard. The use of non-isotopic reporter molecules allows probe detection by both fluorescent and non-fluorescent means. The latter can be carried out more

In situ *hybridization*

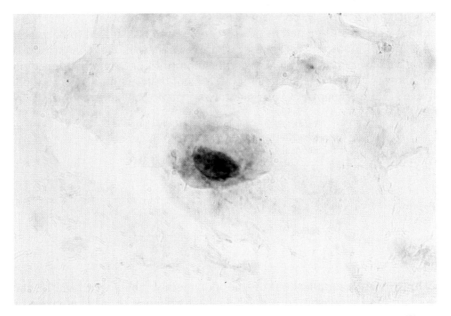

Fig. 12.6. Hybridization of a cervical smear with a cocktail of biotinylated probes for HPV6, 11, 16, 18, 31, 33, 35. Clear nuclear signal can be seen.

quickly than isotopic procedures: are safe; can be manipulated in the context of dual labelling to produce fluorescence signals; and substrate products of contrasting colours detectable with ordinary light.

The simultaneous detection of two nucleic acids can be accomplished in two ways: sequential hybridization and detection of probes labelled with the same reporter molecule, or simultaneous hybridization and detection of probes labelled with different reporters. The former approach is more time-consuming and employs many more practical steps. This leads to higher background noise. Therefore, more than one reporter molecule is required.

The gold standard by which alternative DNA reporters are judged is biotin. This was originally employed in NISH in order to utilize its binding with high affinity to the naturally occurring protein avidin (Langer *et al.* 1981). Many alternatives have been developed and those most frequently used to date have been acetaminofluorene (AAF) and mercury (Hopman *et al.* 1988). These have been detected primarily by fluorescent methods and have been used to visualize chromosome-specific sequences in isolated cells (Cremer *et al.* 1988; Devilee *et al.* 1988; Hopman *et al.* 1988; Trask *et al.* 1988; Nederlof *et al.* 1989). By mixing probes labelled with these reporters and using differential fluorescent detection methods, two, and more recently three (Nederlof *et al.* 1989), chromosomes have been identified within the same cell. However, the safety of these reporter molecules has been questioned because AAF is potently carcinogenic and mercuric cyanide is toxic. In addition, NISH on paraffin sections requires non-fluorescent detection of probe labels because fluorescent methods are less suitable owing to tissue autofluorescence.

Digoxigenin is the aglycone derivative of digoxin and antibodies to digoxin have been shown to cross-react with it with high affinity (Smith *et al.* 1970; Monji *et al.* 1980; Valdes *et al.* 1984). Thus, the well-established antisera used in the assessment and treatment of digoxin toxicity can be applied to the detection of digoxigenin in NISH. Probes labelled with digoxigenin have been shown to be as sensitive as biotin when detected in contrasting colours (Herrington *et al.* 1989*a*) and these two reporters therefore provide suitable alternatives for NISH.

Application to clinical biopsies

The method described in the Appendix allows the simultaneous non-fluorescent detection of two nucleic acids in contrasting colours in clinical biopsies. It is easy, quick, and reliable. The whole procedure, from section cutting to probe visualization, can be completed in approximately 5 hours. This is only 30 minutes longer than the time for detection of one nucleic acid. By mixing probes labelled with digoxigenin and biotin, this technique allows denaturation and hybridization to be performed simultaneously with the target DNA. The detection of biotin using a streptavidin conjugate and of digoxigenin using an alkaline phosphatase conjugate allows mixing of the two detection systems, without cross-reaction, and simultaneous application to the tissue section. The peroxidase substrate (AEC) used to detect biotin-labelled probes, produces a red product and therefore the digoxigenin-labelled probe was detected using a blue substrate, NBT/BCIP (see Appendix) for alkaline phosphatase. Incubation in substrates is performed sequentially, as peroxidase and alkaline phosphatase have different pH optima.

This system was developed for the simultaneous detection of the Y chromosome and HPV (Fig. 12.7) (Herrington *et al.* 1989*a*) and has been applied without modification to the simultaneous detection of two subtypes of HPV (Fig. 12.8) (Herrington *et al.* 1989*b*). It is probable that its application to other nucleic acid sequences that produce NISH signals of different morphology will require the probe-labelling strategy and substrate incubation times to be determined by experiment.

It is important that the two systems used for double labelling are of equivalent sensitivity, particularly if the signals being detected are of similar morphology, in order that the signals generated are directly comparable in all respects other than colour. This has been shown to be the case for the system described here by analysing the frequency distribution (see Fig. 12.9) of the number of dots produced by both systems in CaSki cells (Herrington *et al.* 1989*c*).

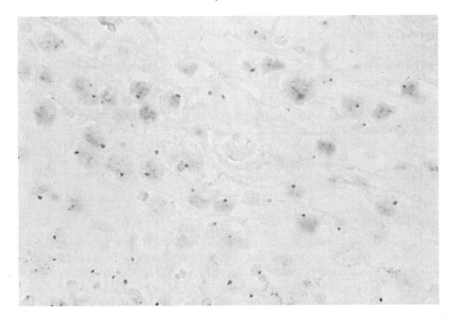

Fig. 12.7. Dual nucleic acid detection: biotinylated HPV6 probe and digoxigenin-labelled Y chromosome-specific probe were hybridized to a section from a male condyloma acuminatum. The HPV signal can be seen as red and the Y signal as blue dots.

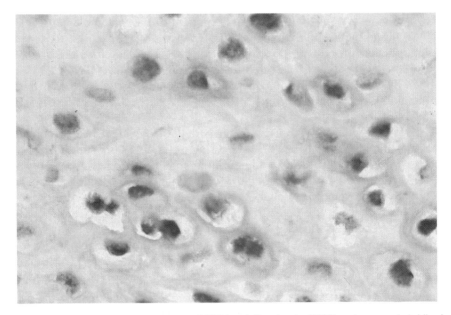

Fig. 12.8. Dual HPV detection: biotinylated HPV6 and digoxigenin HPV11 probes were hybridized to a section from a condyloma acuminatum under conditions of low stringency. The blue signal of digoxigenin HPV11 and the red signal for biotinylated HPV6 can be seen within the same nucleus.

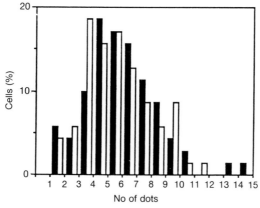

Fig. 12.9. The frequency distribution of the number of dots per cell for the two detection systems used for dual nucleic acid detection are shown. Open columns represent the detection of biotinylated probes with streptavidin peroxidase–AEC and solid columns detection of digoxigenin-labelled probes using NBT–BCIP.

The detection of more than one nucleic acid sequence in the same cell in archival biopsy material has many applications in molecular pathology. The technique described here can be applied to the detection of any two nucleic acid sequences in single cells. For example, the analysis of archival sections for dual viral infections produced no evidence of infection of single cells with HPV6 and 11 (Herrington *et al.* 1989*b*). The analysis of the chromo-some complement of isolated tumour cells (Devilee *et al.* 1988) and of cell samples for prenatal diagnosis (Burns *et al.* 1985; Julien *et al.* 1986; Pinkel *et al.* 1988) has beeen performed using AAF, mercury, and biotin-labelled probes with fluorescent detection. The technique described here will allow these chromosome-specific repeat probes to be applied to archival formalin-fixed, paraffin-embedded material. Although it was developed for the simultaneous detection of DNA sequences, minor modification of this technique will enable its application to the detection of multiple mRNA species in paraffin sections.

Application to cervical smears

As has been mentioned previously, the analysis of cervical smears by NISH is complicated by the lack of duplicated specimens from each patient. In general, this has led to the use of cocktails of probes containing probes for each of the HPV subtypes found in anogenital lesions. Alternatively, cervicovaginal lavage can be used to produce cellular suspensions which can be used to produce multiple smears (Pao *et al.* 1989). However, each smear produced by this method is unlikely to represent an adequate sample of the whole cervix and, more importantly, the collection of such samples is expensive in terms of both time and equipment. By labelling HPV6 and 11 with biotin, and HPV16, 18, 31, 33, and 35 with digoxigenin, the double

(a)

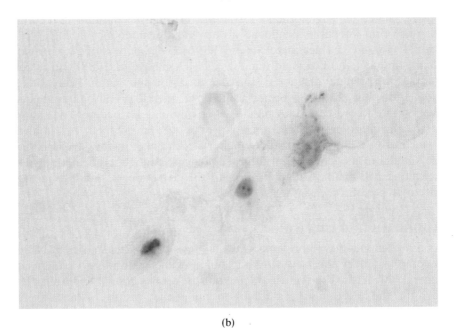

(b)

Fig. 12.10. The application of the dual detection system to cervical smears produces a red signal (a) indicative of HPV6/11 infection and a blue signal (b) indicative of HPV16/18/31/33/35 infection (see text).

detection system described for archival biopsies can be applied to cervical smears. Preliminary results suggest that this method will be routinely applicable and easily interpretable, with a red signal indicative of 'benign' infection (Fig. 12.10(a)) and a blue signal of 'malignant' infection (Fig. 12.10(b)). This approach is still being evaluated but may provide prognostic information relevant to individual patients based on the analysis of a single cervical smear.

Conclusion

In situ hybridization can be used to analyse the presence or absence of nucleic acids in a variety of cell and tissue samples. *In situ* hybridization is the only means by which a definite diagnosis of viral infection can be made in conjunction with the analysis of the morphology of the infected tissue. The methods described here are easy and quick and can be applied to the rapid screening of archival and current tissue blocks, thus providing a wealth of epidemiological data about oncogenic infectious disease. Although the sensitivity of methods routinely available is at present limited, it is likely that low-copy-number sequences will soon be detectable in archival biopsies. The simultaneous detection of two nucleic acids on the same section and in the same nucleus in archival samples has allowed the investigation of double viral infection. Its application to other areas, for example the field of interphase cytogenetics, will allow the investigation of the relationship between nucleic acids in individual nuclei in archival material. Similarly, the application of these systems to the analysis of cervical smears will allow more effective HPV typing together with correlation with morphology.

References

Bancroft, J. D. and Stevens, A. (eds) (1977). *Theory and practice of histological techniques.* Churchill Livingstone, Edinburgh.

Beckmann, A. M., Myerson, D., Daling, J. R., Kiviat, N. B., Fenoglio, C. M., and McDougall, J. K. (1985). Detection and localisation of human papillomavirus DNA in human genital condylomas by *in situ* hybridisation with biotinylated probes. *J. Med. Virol.*, **16**, 265–73.

Bhatt, B., Burns, J., Flannery, D., and McGee, J. O'D. (1988). Direct visualisation of single copy genes on banded metaphase chromosomes by nonisotopic in situ hybridisation. *Nucleic Acids Res.*, **16**, 3951–61.

Blum, H. E., Stowring, L., Figus, A., Montgomery, C. K., Haase, A. T., and Vyas, G. N. (1983). Detection of hepatitis B virus DNA in hepatocytes, bile duct epithelium and vascular elements by *in situ* hybridisation. *Proc. Natl Acad. Sci.*, **80**, 6685–8.

Boerman, R. H., Arnolous, E. P. J., Raap, A. K., Peters, A. C. B., Ter Schegget, J., and Van Der Ploeg, M. (1989). Diagnosis of progressive multifocal leucoencephalopathy by hybridisation techniques. *J. Clin. Pathol.*, **42**, 153–61.

Bonfiglio, T. A. (1988). Human papillomaviruses and cancer of the uterine cervix. *Human Pathol.*, **19**, 621–2.

Burk, R. D., Kadish, A. S., Calderin, S., and Romney, S. L. (1986). Human papillomavirus infection of the cervix detected by cervicovaginal lavage and molecular hybridisation: correlation with biopsy results and Papanicolau smears. *Am. J. Obstet. Gynaecol.*, **154**, 982–9.

Burns, J., Chan, V. T.-W., Jonasson, J. A., Fleming, K. A., Taylor, S., and McGee, J. O'D. (1985). Sensitive system for visualising biotinylated DNA probes hybridised *in situ*: rapid sex determination in intact cells. *J. Clin. Pathol.*, **38**, 1085–92.

Burns, J., Redfern, D. R. M., Esiri, M. M., and McGee, J. O'D. (1986). Human and viral gene detection in routine paraffin embedded tissue by *in situ* hybridisation with biotinylated probes: viral localisation in herpes encephalitis. *J. Clin. Pathol.*, **39**, 1066–73.

Burns, J., Graham, A. K., Frank, C., Fleming, K. A., Evans, M. F., and McGee, J. O'D. (1987). Detection of low copy human papilloma virus DNA and mRNA in routine paraffin sections of cervix by non-isotopic *in situ* hybridisation. *J. Clin. Pathol.*, **40**, 858–64.

Burns, J., Graham, A. K., and McGee, J. O'D. (1988). Non-isotopic detection of *in situ* nucleic acid in cervix: an updated protocol. *J. Clin. Pathol.*, **41**, 897–9.

Cordell, J. L., Falini, B., Erber, W. N., Ghosh, A. K., Abdulaziz, Z., MacDonald, S., Pulford, K. A. F., Stein, H., and Mason, D. Y. (1984). Immunoenzymatic labelling of monoclonal antibodies using immune complexes of alkaline phosphatase and monoclonal anti-alkaline phosphatase (APAAP complexes). *J. Histochem. Cytochem.*, **32**, 219–29.

Cremer, T., Tesin, D., Hopman, A. H. N., and Manuelidis, L. (1988). Rapid interphase and metaphase assessment of specific chromosomal changes in neuroectodermal cells by *in situ* hybridisation with chemically modified DNA probes. *Exp. Cell. Res.*, **176**, 199–220.

De Villiers, E-M., Schneider, A., Miklaw, H., Papendick, U., Wagner, D., Wesch, H., Wahrendorf, J., and Zur Hansen, H. (1987). Human papillomavirus infections in women with and without abnormal cervical cytology. *Lancet*, **ii**, 703–6.

Devilee, P., Thierry, R. F., Kievits, T., Kolluri, R., Hopman, A. H. N., Willard, H. F., Pearson, P. L., and Cornelisse, C. J. (1988). Detection of chromosome aneuploidy in interphase nuclei from human primary breast tumours using chromosome-specific repetitive DNA probes. *Cancer Res.*, **48**, 5825–30.

Dirks, R. W., Raap, A. K., Van Minnen, J., Vreugdenhil, E., Smit, A. B., and Van Der Ploeg, M. (1989). Detection of mRNA molecules coding for neuropeptide hormones of the pond snail *Lymnaea stagnalis* by radioactive and non radioactive *in situ* hybridisation: a model study for mRNA detection. *J. Histochem. Cytochem.*, **37**, 7–14.

Grody, W. W., Cheng, L., and Lewin, K. J. (1987). *In situ* viral DNA hybridisation in diagnostic surgical pathology. *Human Pathol.*, **18**, 535–43.

Haase, A. T. (1986). Analysis of viral infections by *in situ* hybridization. *J. Histochem. Cytochem.*, **34**, 27–32.

Haase, A. T., Gantz, D., Eble, B., Walker, D., Stowring, L., Ventura, P., Blum, H., Wietgrefe, S., Zupancic, M., Tourtellotte, W., Gibbs, C. J., Norrby, E., and Rozenblatt, S. (1985*a*). Natural history of restricted synthesis and expression of measles virus genes in subacute sclerosing panencephalitis. *Proc. Natl Acad. Sci.*, **82**, 3020–4.

Haase, A. T., Walker, D., Ventura, P., Geballe, A., Blum, H., Brahic, M., Goldberg, R., and O'Brien, K. (1985*b*). Detection of two viral genomes in single cells by double-label hybridisation *in situ* and color microradioautography. *Science*, **227**, 189–92.

Herrington, C. S., Burns, J., Graham, A. K., Bhatt, B., and McGee, J. O'D. (1989*a*). Interphase cytogenetics using biotin and digoxigenin labelled probes II: simultaneous detection of two nucleic acid species in individual nuclei. *J. Clin. Pathol.*, **42**, 601–6.

Herrington, C. S., Burns, J., Graham, A. K., and McGee, J. O'D. (1989*b*). Evaluation of discriminative stringency conditions for the distinction of closely homologous DNA sequences in clinical biopsies. (Submitted.)

Herrington, C. S., Burns, J., Graham, A. K., Evans, M. F., and McGee, J. O'D. (1989*c*). Interphase cytogenetics using biotin and digoxigenin labelled probes I: relative sensitivity of both reporters for detection of HPV16 in CaSki cells. *J. Clin. Pathol.*, **42**, 592–600.

Hopman, A. H. N., Ramaekers, F. C. S., Raap, A. K., Beck, J. L. M., Devilee, P., Van Der Ploeg, M., and Vooijs, G. P. (1988). *In situ* hybridisation as a tool to study numerical chromosome aberrations in solid bladder tumours. *Histochemistry*, **89**, 307–16.

Hsu, S.-M., Raine, L., and Fanger, H. (1981). Use of avidin-biotin-peroxidase complex (ABC) in immunoperoxidase techniques. *J. Histochem. Cytochem.*, **29**, 577–80.

Hsu, S.-M. and Soban, E. (1982). Color modification of diaminobenzidine (DAB) precipitation by metallic ions and its application for double immunohistochemistry. *J. Histochem. Cytochem.*, **30**, 1079–82.

Julien, C., Bazin, A., Guyot, B., Forestier, F., and Daffos, F. (1986). Rapid prenatal diagnosis of Down's syndrome with *in situ* hybridisation of fluorescent DNA probes. *Lancet*, **2**, 863–4.

Langer, P. R., Waldrop, A. A., and Ward, D. C. (1981). Enzymatic synthesis of biotin-labeled polynucleotides: novel nucleic acid affinity probes. *Proc. Natl Acad. Sci.*, **78**, 6633–7.

Lorincz, A. T., Temple, G. F., Patterson, J. A., Jenson, A. B., Kurman, R. J., and Lancaster, W. D. (1986). Correlation of cellular atypia and human papillomavirus deoxyribonucleic acid sequences in exfoliated cells of the uterine cervix. *Obstet. Gynaecol.*, **68**, 508–12.

Matthews, J. A. and Kricka, L. J. (1988). Analytical strategies for the use of DNA probes. *Analyt. Biochem.*, **169**, 1–25.

McDougall, J. K. (1988). *In situ* hybridisation for viral gene detection. *ISI Atlas of biochemistry*, pp. 6–10.

Melchers, W. J. G., Herbrink, P., Walboomers, J. M. M., Meijer, C. J. L. M., Drift, Hvd., Lindeman, J., and Quint, W. G. V. (1989). Optimisation of human papillomavirus genotype determination in cervical scrapes by a modified filter in situ hybridisation test. *J. Clin. Microbiol*, **27**, 106–10.

Mincheva, A., Gissman, L., and Zur Hausen, H. (1987). Chromosomal integration sites of human papillomavirus DNA in three cervical cancer cell lines mapped by *in situ* hybridisation. *Med. Microbiol. Immunol.,* **176**, 245–56.

Monji, N., Ali, H., and Castro, A. (1980). Quantification of digoxin by enzyme immunoassay: synthesis of a maleimide derivative of digoxigenin succinate for enzyme coupling. *Experientia,* **36**, 1141–3.

Nederlof, P. M., Robinson, D., Abuknesha, R., Wiegant, J., Hopman, A. H. N., Tanke, H. J., and Raap, A. K. (1989). Three-color fluorescence *in situ* hybridisation for the simultaneous detection of multiple nucleic acid sequences. *Cytometry,* **10**, 20–7.

Pao, C. C., Lai, C.-H., Wu, S.-Y., Young, K.-C., Chang, P.-L., and Soong, Y.-K. (1989). Detection of human papillomaviruses in exfoliated cervicovaginal cells by *in situ* DNA hybridisation analysis. *J. Clin. Microbiol.,* **27**, 168–73.

Pezzella, M., Pezzella, F., Galli, C., Macchi, B., Verani, P., Sorice, F., and Baroni, C. D. (1987). In situ hybridisation of human immunodeficiency virus (HTLV-III) in cryostat sections of lymph nodes of lymphadenopathy syndrome patients. *J. Med. Virol.,* **22**, 135–42.

Pfister, H. and Fuchs, P. G. (1987). In *Papillomaviruses and human disease* (ed. K. Syrjanen, L. Gissman, and L. G. Koss), pp. 1–18. Springer-Verlag, New York.

Pinkel, D. *et al.* (1988). Fluorescence *in situ* hybridisation with human chromosome specific libraries: Detection of trisomy 21 and translocations of chromosome 4. *Proc. Natl Acad. Sci.,* **85**, 9138–47.

Popescu, N. C., Amsbaugh, S. C., and DiPaolo, J. A. (1987*a*). HPV type 18 DNA is integrated at a single chromosome site in cervical carcinoma cell line SW756. *J. Virol.,* **51**, 1682–5.

Popescu, N. C., DiPaolo, J. A., and Amsbaugh, S. C. (1987*b*). Integration sites of HPV18 DNA sequences on HeLa cell chromosomes. *Cytogenet. Cell Genet.,* **44**, 58–62.

Porter, H., Quantrill, A. M., and Fleming, K. A. (1988). B19 parvovirus infection of myocardial cells. *Lancet,* **1**, 535–6.

Saiki, R. K., Scharf, S., Faloona, F., Mullis, K. B., Horn, G. T., Ehrlich, H. A., and Arnheim, N. (1985). Enzymatic amplification of β-globin genomic sequences and restriction site analysis for diagnosis of sickle cell anaemia. *Science,* **230**, 1350–4.

Shibata, D. K., Arnheim, M., and Martin, W. J. (1988). Detection of human papillomavirus in paraffin embedded tissue using the polymerase chain reaction. *J. Exp. Med.,* **167**, 225–30.

Smith, T. W., Butler, V., and Haber, E. (1970). Characterisation of antibodies of high affinity and specificity for the digitalis glycoside digoxin. *Biochemistry,* **9**, 331–7.

Sternberger, L. A., Hardy, P. H., Cuculis, J. J., and Meyer, H. G. (1970). The unlabelled antibody-enzyme method of immunohistochemistry. Preparation and properties of soluble antigen-antibody complex (horseradish peroxidase-anti-horseradish peroxidase) and its use in identification of spirochetes. *J. Histochem. Cytochem.,* **18**, 315–33.

Syrjanen, K. J. (1987). Biology of HPV infections and their role in squamous cell carcinogenesis. *Med. Biol.,* **65**, 21–39.

Syrjanen, S., Partanen, P., Mantyjarvi, R., and Syrjanen, K. (1988). Sensitivity of *in situ* hybridisation techniques using biotin and ^{35}S labelled human papillomavirus (HPV) DNA probes. *J. Virol. Methods,* **19**, 225–38.

Syrjanen, K., Syrjanen, S., Kellokoski, J., Karja, J., and Mantyjarvi, R. (1989). Human papillomavirus (HPV) type 6 and 16 DNA sequences in bronchial squamous cell carcinomas demonstrated by *in situ* hybridisation. *Lung,* **16**, 33–42.

Taussig, M. J. (1984). *Processes in Pathology* (2nd edn). Blackwell Scientific, Oxford.

Toon, P. G., Arrand, J. R., Wilson, L. P., and Sharp, D. S. (1986). Human papillomavirus infection of the uterine cervix of women without cytological signs of neoplasia. *Br. Med. J.,* **293**, 1261–4.

Trask, B., van den Eugh, G., Pinkel, D., Mullikin, J., Waldman, F., van Dekken, H., and Gray, J. (1988). Fluorescence *in situ* hybridisation to interphase nuclei in suspension allows flow cytometric analysis of chromosome content and microscopic analysis of nuclear organisation. *Human Genet.,* **78**, 251–9.

Valdes, R., Brown, B. A., and Graves, S. W. (1984). Variable cross-reactivity of digoxin metabolites in digoxin immunoassays. *Am. J. Clin. Pathol.,* **82**, 210–13.

Wagner, D., Ikenberg, H., Boehm, N., and Gissmann, L. (1984). Identification of human papillomavirus in cervical swabs by deoxyribonucleic acid *in situ* hybridisation. *Obstet. Gynaecol.,* **64**, 767–72.

Wagner, L., and Worman, C. P. (1988). Colour-contrast staining of two different lymphocyte sub-populations: a two-colour modification of alkaline phosphatase monoclonal anti-alkaline phosphatase complex technique. *Stain technol.,* **63**, 129–35.

Walboomers, J. M. M. *et al.* (1988). Sensitivity of *in situ* detection with biotinylated probes of human papilloma virus type 16 DNA in frozen tissue sections of squamous cell carcinomas of the cervix. *Am. J. Pathol.,* **131**, 587–94.

Warford, A. (1988). *In situ* hybridisation: a new tool in pathology. *Med. Lab. Sci.,* **45**, 381–94.

Wells, M., Griffiths, S., Lewis, F., and Bird, C. C. (1987). Demonstration of human papillomavirus types in paraffin processed tissue from human anogenital lesions by *in situ* DNA hybridisation. *J. Pathol.,* **152**, 77–82.

Wolber, R. A. and Lloyd, R. V. (1988). Cytomegalovirus detection by nonisotopic *in situ* DNA hybridisation and viral antigen immunostaining using a two-colour technique. *Human Pathol.,* **19**, 736–41.

Young, L. S., Bevan, I. S., Johnson, M. A., Blomfield, P. I., Bromidge, T., Maitland, N. J., and Woodman, C. B. J. (1988). The polymerase chain reaction: a new epidemiological tool for investigating cervical human papillomavirus infection. *Br. Med. J.,* **298**, 14–18.

Appendix 12.1

Materials

All organic chemicals, unless otherwise stated, were purchased from BDH, UK.

1. Digoxigenin dUTP and alkaline phosphatase conjugated anti-digoxigenin (Boehringer Mannheim, FRG; from kit no. 1093 657)

2. Biotin-7-dATP and dNTP mixture for biotinylation (Gibco/BRL)

3. DNase 1 (Worthingtons, UK; Cat. No. L500 06330)

4. DNA polymerase 1 (Boerhinger Mannheim, FRG; Cat. No. 642 711)

5. Glycogen (Boerhinger Mannheim, FRG; Cat. No. 901 393)

6. Tris-EDTA (TE)—10 mM Tris-HCl, 1 mM EDTA, pH 8

7. Four-spot multiwell slides (Hendley, Essex: spot diameter 12 mm)

8. Decon 90 (Philip Harris, UK; Cat. No. D35-460)

9. Aminopropyltriethoxysilane (silane) (Sigma, UK; Cat. No. A3648)

10. Proteinase K (Boerhinger Mannheim, FRG; Cat. No. 745 723)

11. Phosphate-buffered saline (PBS)—10 mM phosphate, 150 mM NaCl, pH 7.4

12. Formamide (Sigma, UK: Cat. No. F7503)

13. Dextran sulphate (Sigma, UK; Cat. No. D6001)

14. Standard saline citrate (SSC)—1 × SSC: 150 mM NaCl, 15 mM sodium citrate

15. Coverslips (Chance Raymond Lamb, UK); Cat. No. E/23—22 × 64 mm and 14 mm^2

16. Terasaki plates (Gibco/Nunc, UK; Cat. No. 1-36528A)

17. Tris-buffered saline (TBS)—50 mM Tris-HCl, 100 mM NaCl, pH 7.2

18. Bovine serum albumin (Sigma, UK; Cat. No. B2518)

19. Triton-x-100 (Sigma, UK; Cat. No. T8761)

20. Streptavidin-peroxidase (Dako, UK; Cat. No. P397)

21. Avidin alkaline phosphatase (Dako, UK; Cat. No. D365)

22. Monoclonal anti-biotin (Dako, UK; Cat. No. M743)

23. Biotinylated rabbit anti-mouse [F(ab')$_2$ fragment] (Dako, UK; Cat. No. E413)

21. Substrate buffer—50 mM Tris-HCl, 100 mM NaCl, 1 mM MgCl$_2$, pH 9.5

22. AEC. 3-amino-9-ethylcarbazole (Sigma, UK; Cat. No. A5754)

23. NBT. Nitroblue tetrazolium (Sigma, UK; Cat. No. N6876)

24. BCIP. 5-bromo-4-chloro-3-indolylphosphate (Sigma, UK; Cat. No. B8503)

Methods
Probe labelling
1. Mix:

	Digoxigenin	Biotin
Plasmid DNA	1 μg	1 μg
Nucleotide Mix*	1 μl	5 μl
Biotin-dATP	—	2.5 μl
DNase 1†	1–5 ng	1–5 ng

   * For biotin labelling, the nucleotide mix contains 0.2 mM of each of dCTP, dTTP, and dGTP. This is made up from individual nucleotides. For digoxigenin labelling, the mix contains all nucleotides including digoxigenin-dUTP/dTTP at a concentration of 1 mM.
   † The amount of DNase required is assessed by alkaline gel electrophoresis followed by Southern transfer detection as described previously. DNase is adjusted to give a median probe size of 200–400 bp.

2. Add distilled water (dH$_2$O) to 49 μl and incubate at 37 °C for 2 min.
3. Add 1 μl DNA polymerase 1 (5 U/μl) and incubate at 14 °C for 150 min.
4. Add 5 μl 200 mM EDTA and heat at 75 °C for 10 min to terminate reaction.
5. Add 1 μl mussel glycogen (20 mg/ml)
6. Add 1/10 vol. 3M NaOAc, pH 6, then 2 vols. absolute ethanol (EtOH).
7. Incubate on dry ice for 20 min or at − 20 °C overnight.
8. Spin at 12,000 rpm for 20 min.
9. Wash in 80% EtOH × 4.
10. Lyophilize and resuspend in 50 μl TE.
11. Store labelled probes at either − 20 °C or 4 °C. If stored at − 20 °C, do not repeatedly freeze–thaw.

This procedure can be scaled up by multiplication of the reaction volumes.

Preparation of slides
1. Clean four-spot multiwell slides (12 mm spot diameter) in 2% Decon 90 at 60 °C for 30 min.
2. Rinse in dH$_2$O then acetone, and air dry.
3. Immerse in 2% aminopropylethoxysilane in acetone for 30 min.
4. Wash in dH$_2$O and air dry at 37 °C.

5. Slides prepared in this way can be stored indefinitely at room temperature.

Preparation of sections

1. Cut 5 μm sections from routine paraffin embedded blocks onto slides prepared as above.
2. Bake sections either (a) overnight at 60 °C or (b) for 45 min at 75 °C.
3 Store slides at room temperature.

Dewaxing of sections

1. Heat to 75 °C for 15 min.
2. Plunge into xylene and wash for 10 min. Change xylene once.
3. Clear in 99% ethanol (industrial grade).
4. Wash in dH$_2$O.

Unmasking of nucleic acids

1. Store proteinase K at 20 mg/ml in dH$_2$O at -20 °C.
2. Dilute to 500 μg/ml in PBS.
3. Spot onto slides, place in Terasaki plates, and float in water bath at 37 °C for 15 min.
4. Wash in dH$_2$O and air dry at 75 °C.

Preparation of hybridization mixture (HM)

1. Mix: Deionized formamide   5 ml
          50% (w/v) dextran sulphate 1 ml
          20 × SSC   1 ml
2. Adjust pH to 7.0 and store at 4 °C.

Application of probe to section

For each spot:

1. Add 1 μl (10 ng) of each of biotin and/or digoxigenin labelled probes to 7 μl of HM.
2. Add TE to make 10 μl total volume.
3. Spot 8 μl of the mixture on to each well.
4. Cover each spot with 14 mm coverslip. The latter need not be silanized and should not be sealed with rubber solution as is frequently recommended.

Denaturation/hybridization

1. Place slides in Terasaki plates (2 slides per plate).
2. Denture on a hot plate in hot air oven at 95 °C for 15 min.

3. Hybridize at 42 °C in hot air oven for 2 h.

Stringency washing and blocking procedure

1. Wash slides in 4 × SSC at room temperature twice—5 min each.
2. Wash in appropriate solution for high stringency* e.g. 50% formamide/ 0.1 × SSC (if required for discriminating closely homologous sequences).
3. Wash in 4 × SSC at room temperature for 5 min.
4. Incubate for 30 min in blocking solution (TBT)—TBS containing 5% (w/v) bovine serum albumin, 0.05% (v/v) Triton-x-100.

Detection of biotinylated probes

(a) Conventional signal detection

1. Incubate at 22 °C for 30 min in avidin alkaline phosphatase diluted 1:100 in TBT.
2. Remove unbound conjugate by washing twice in TBS for 5 min.
3. Incubate in nitroblue tetrazolium (NBT)/5-bromo-4-chloro-3-indolyl-phosphate (BCIP) for 15–20 min.
4. Terminate the reaction by washing in distilled water for 5 min.
5. Air dry the slides at 42 °C and mount in glycerol jelly.

(b) Amplified signal detection

1. Incubate at 22 °C for 30 min in monoclonal anti-biotin diluted 1:50 in TBT.
2. Wash in TBS for 5 min twice.
3. Incubate at 22 °C for 30 min in biotinylated rabbit anti-mouse [F(ab')$_2$ fragment] diluted 1:200 in TBT.
4. Wash in TBS for 5 min twice.
5. Incubate in avidin alkaline phosphatase diluted 1:50 in TBT containing 5% non-fat milk.
6. Follow steps 2–5 of the conventional detection system.

Double probe detection

1. Incubate at 22 °C for 30 min in a mixture of streptavidin-peroxidase conjugate diluted 1:100 in TBT and alkaline phosphatase conjugated anti-digoxigen diluted 1:600 in TBT; in practice, 1 μl of antibody and 6 μl of streptavidin is added to 600 μl of TBT.
2. Remove unbound conjugate by washing twice in 50 mM Tris-HCl 100 mM NaCl pH 7.2 (TBS) for 5 min.

---

* Adjust all washing solutions to pH 7 with 5 M HCl. The temperature of the solution should be monitored directly using a mercury thermometer. Washing should be carried out for 30 min.

3. Incubate in the peroxidase substrate, 3-amino-9-ethylcarbazole (AEC) for 30 min at 22 °C.
4. Terminate the reaction by thorough washing in TBS.
5. Wash in substrate buffer for 5 min at 22 °C.
6. Incubate in NBT/BCIP for 20–40 min.
7. Terminate the reaction by washing in distilled water for 5 min.
8. Air dry the slides at 42 °C and mount in glycerol jelly.

Substrate preparation

(a) NBT/BCIP

1. Dissolve 10 mg NBT in 200 μl dimethylformamide (DMF).
2. Add 1 ml substrate buffer at 37 °C and add mixture dropwise to 30 ml substrate buffer at 37 °C.
3. Dissolve 5 mg BCIP in 200 μl DMF and add slowly to the above mixture.
4. Aliquot and store at -20 °C.

(b) AEC

1. Dissolve 2 mg of AEC in 1.2 ml dimethylsulphoxide in a glass tube.
2. Add to 10 ml 20 mM acetate buffer, pH 5.0–5.2.
3. Add 0.8 μl of 30% (v/v) hydrogen peroxide and use the final mix immediately.

Glycerol jelly (Bancroft and Stevens 1977)

1. Dissolve 10 g gelatine in 60 ml dH_2O in 37 °C waterbath.
2. Add 70 ml glycerol and 0.25 g phenol.
3. Store at room temperature (as a gel) or in 42 °C oven (when it is a liquid).

Note added in proof

A recent study of 210 cases has confirmed the high positivity rate of normal cervical smears using PCR (Tidy *et al.* 1989*a*), with 84 per cent containing HPV16 sequences. Only 70 per cent of dyskaryotic but 100 per cent of carcinomata were positive for HPV16. The analysis of the PCR reaction products produced from these cases has identified 2 subtypes of HPV, namely HPV16a and b (Tidy *et al.* 1989*b*). HPV16b has a 21 bp deletion compared with HPV16a and is associated with normal cytology. HPV16a segregates with abnormal cytology. Thus, the oncogenic potential of HPVs may be associated with a short deletion in the upstream regulatory region of the viral genome.

References

Tidy, J. A., Parry, G. C. N., Ward, P., Coleman, D. V., Peto, J., Malcolm, A. D. B., and Farrell, P. J. (1989*a*). High rate of human papillomavirus type 16 infection in cytologically normal cervices. *Lancet*, **i**, 434.

Tidy, J. A., Vousden, K. H., and Farrell, P. J. (1989*b*). Relation between infection with a subtype of HPV16 and cervical neoplasia. *Lancet*, **i**, 1225–7.

13

Photomicrography

SAVILE BRADBURY

Introduction

Photomicrographs are constantly required for illustrating routine reports and publications, as well as for illustrating lectures. The ability to take good photomicrographs might, at first sight, not be expected from practising morphologists. It could be argued that they do not have the specialist knowledge of equipment and that this facility is often offered by staff photographers in research laboratories or in hospital departments of medical illustration. Using such services, however, often requires taking the sections to a studio separate from the laboratory, with the consequent inconvenience that this causes. There may be difficulties in relocating the exact areas of interest in a specimen. The photographic department may have many other photographic jobs outstanding at the time that the photomicrographs are required, so that long delays may ensue before the material can be photographed. Again, a specialist service may not be available when it is most required (e.g. at weekends or after regular working hours). For these reasons, the ability to take high-quality photomicrographs should be regarded as an essential skill for any practising laboratory scientist. This is especially true for scientists working in the field of *in situ* hybridization. The preparative techniques are very laborious and it is important to be able to produce convincing illustrations of the end result. This is often difficult, in particular since the contrast may be poor. Some of these problems may be solved by the traditional techniques used for the photography of autoradiographs, especially by the use of dark-ground illumination.

With the high-quality equipment now available from most manufacturers and with modern photographic materials, it is relatively easy to produce excellent photomicrographs both in colour and in black and white. It is not necessary to have a fully equipped darkroom, as in the majority of cases the actual printing of the negatives would be passed on to the departmental photographers. The processing of monochrome film, however, is best carried out by the person actually taking the photomicrographs, so that if for any reason the negatives are unsatisfactory the exposures may be repeated without undue delay. This implies that the facilities for developing, fixing, and washing monochrome films should be available in every laboratory where micrographs are produced. Details of the equipment required and techniques are given in the Appendix to this chapter. The critical steps in producing a good photomicrograph may be summarized under the following heads:

(1) choice and preparation of the specimen;

(2) selection of suitable microscope optics and their correct use;

(3) obtaining correct photographic focus, exposure, and contrast.

Aspects of specimen preparation will not be discussed further in this chapter as they vary greatly according to the especial needs of individual scientists.

Microscopical aspects

A thorough understanding of the operating principles of the microscope is essential as a prelude to taking good photomicrographs. Without such an understanding, it is very easy for spurious images to result. Such optical artefacts may arise from the use of inadequate, unsuitable, or defective equipment. More often it is found that images of poor quality are the result of operator error or ignorance, when microscopes of good quality are incorrectly set up or are in poor adjustment. Good microscopy is, therefore, essential if the researcher is to record an image that represents as faithfully as possible the interactions between the light and his specimen.

It is not possible here to enter into an exhaustive discussion of the details of microscopy nor to consider the theory behind its use. The interested reader is referred to one of the texts on microscopy (e.g. Bradbury 1989; Delly 1988; Hartley 1979; Spencer 1982).

The objective and eyepiece

The correct selection of the objective and eyepiece combination, together with the correct use of the substage condenser and the illuminating system, are vital controlling factors for the three major requirements of good microscopy. These are

(1) achieving resolving power (the ability to make fine detail in an object distinguishable in an image;

(2) achieving visibility or contrast;

(3) achieving a suitable magnification.

The first part of the imaging system of the microscope is the objective. It forms the real, magnified primary imge in which all of the fine detail of the object should be resolved. It is thus clear that the correct choice of objective is essential for good photomicrography. When deciding which objective to use, it is usual to consider the numerical aperture (NA), the initial magnification, and the correction category.

The numerical aperture is a parameter that is an expression not only of
the light-gathering capability of the objective but also of its ability to
resolve fine detail. In general, the greater the value of the NA, then the
better the detail in the image and the brighter this will be, a factor of
importance when high magnifications are in use or when techniques such as
phase-contrast or fluorescence are employed. The values of the numerical
aperture for some representative objectives are listed in Table 13.1. It
should be noted that in general the higher the initial magnification, the
higher the NA, although special objectives designed for fluorescence may
have very high numerical apertures and relatively low initial magnifications.
Objectives that are used 'dry' (that is, with air intervening between the front
lens of the objective and the cover of the specimen) are limited to a
maximum NA of 0.95. If higher values are required, it is necessary to
immerse the objective by adding an optically homogeneous medium
(usually a synthetic oil) to connect the front of the objective with the cover
glass.

Objectives are classified according to their optical corrections. The most
basic are the so-called achromats, in which chromatic aberration is
minimized for two wavelengths (one in the red and one in the blue) and
spherical aberration is minimized for one wavelength, usually in the green
region of the spectrum. These perform well for routine observations but are

Table 13.1. Magnification and numerical apertures of
some typical microscope objectives

Objective type	Magnification	Numerical aperture
Achromat	4	0.10
Achromat	10	0.25
Achromat	20	0.40
Achromat	40	0.65
Achromat*	100	1.25
Semi-apochromat	1	0.04
Semi-apochromat	2	0.08
Semi-apochromat	4	0.13
Semi-apochromat	10	0.30
Semi-apochromat	20	0.46
Semi-apochromat	40	0.70
Semi-apochromat*	100	1.25
Apochromat	4	0.16
Apochromat	10	0.40
Apochromat	20	0.70
Apochromat†	40	0.95
Apochromat*	100	1.35

* Oil-immersion objective.
† With a correction collar.
For an · objective of any given magnification, note the increase in
numerical aperture from achromat through semi-apochromat to apo-
chromat.

only really suitable for photomicrography if used with monochromatic light (as, for example, when using phase contrast). The most highly corrected objectives are the apochromats, which not only have an increased numerical aperture (when compared to achromats of the same initial magnification), but have chromatic aberration minimized at three wavelengths (in the red, green, and blue), the sine condition fulfilled for two wavelengths, and spherical aberration minimized for a single wavelength (as in achromats). These objectives are capable of giving images of the very highest quality, provided that they are used correctly and that the specimen is suited to their use. Apochromats often have very short free working distances; this means that specimens mounted beneath thick cover glasses are often incapable of being focused by these lenses. Also, high-aperture 'dry' apochromats are very sensitive to coverglass thickness and will only produce a well-corrected image if used with a cover of the thickness for which they were designed. If the thickness is incorrect, they produce an image marred by marked spherical aberration, which causes unsharpness. In order to overcome this, many apochromats are provided with a rotatable correction collar that allows the user to adjust them in order to compensate for spherical aberration introduced by the specimen or mount. It is often the case, however, that such collars are incorrectly used, with consequent unsharpness in the images that may lead to considerable (and unjustified) dissatisfaction with the objective. If an immersion lens is used, however, then aberrations introduced from the use of a cover glass of incorrect thickness are of much less significance.

Perhaps the best compromise for photomicrography is to use objectives with an intermediate degree of correction between achromats and apochromats. Such objectives, often termed 'semi-apochromats' are available from all manufacturers, often under a specific trade name. They are easier to use, as they are more tolerant of incorrect cover thickness and of operator error. They produce images that in practice are almost impossible to distinguish from those provided by the very much more expensive apochromats.

All objectives, whatever their correction class, are now available in a special form computed to give as flat an image field as possible. Such objectives (flat-field or 'Plan' objectives) are essential if good photomicrographs are to be obtained. The eye is very intolerant of peripheral unsharpness due to field curvature in a photograph, whereas in actual observation at the instrument such unsharpness is constantly being corrected by the use of the fine focus.

Use of the correct eyepiece is also important, as this component is often required to complete the optical corrections provided by the objective. In practice, this means that the eyepiece in a photomicrographic system should be that recommended by the maker. Eyepieces are available in several different magnifications in order to complete the total magnification provided by the system. As a rule it should be remembered that it is good practice to use the objective with the highest possible NA that provides the requisite magnification (when used with an eyepiece providing further magnification of × 6 or at the most × 10). Figure 13.1 shows the effect of obtaining the same magnification in two ways. Micrograph A was produced by the use of an objective with a high NA and a high initial magnification

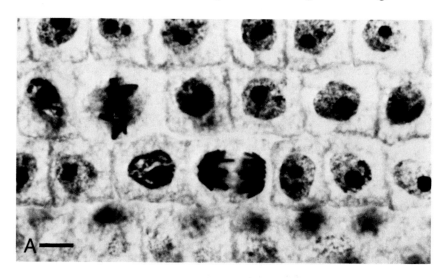

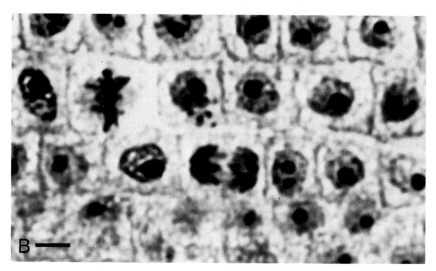

Fig. 13.1. (A) Micrograph of mitotic figures in a section of onion root tip stained with iron haematoxylin. Photographed with objective of NA 1.3 and a low-power eyepiece. Note the sharp detail visible in the image. (B) The same field photographed using an objective of NA 0.3 and a high-power eyepiece to obtain the same magnification. Note the overall lack of sharpness in the image. The bar represents 10 μm.

used with a low-power eyepiece. It is sharper (i.e. it has a greater resolution) than micrograph B. The latter was produced with an objective of low NA (and therefore of low resolving power) and low initial magnification. For this micrograph the requisite magnification was produced largely by the high-power eyepiece; it is evident that this combination gives an image of very poor quality indeed.

The condenser and illuminating system

One of the prime functions of the substage condenser is to deliver a cone of light of controlled numerical aperture to the objective. In order to obtain the best resolution (assuming a contrasty specimen) it is important to make sure that the condenser is filling at least 90 per cent of the numerical aperture of the objective. It is also important in order to avoid excessive glare to make sure that the condenser is illuminating only that area of the specimen that is under actual observation at any time. These conditions are more easily achieved if a well-corrected achromatic–aplanatic condenser is used rather than with the poorly-corrected Abbe illuminator that is so often fitted to routine laboratory instruments. If extreme resolution is sought with the very highest power oil-immersion objective, then it is desirable to use the condenser in oil immersion contact with the under surface of the slide. This will allow a cone of light of NA greater than 1 to be produced, so filling the aperture of the objective more completely than would be possible if the condenser were used 'dry'. In use the condenser must be

(1) focused on the plane of the object;

(2) correctly centred so that its optical axis coincides with that of the microscope objective; and

(3) have the aperture diaphragm in its front focal plane set so that the condenser delivers a cone of light of adequate aperture.

This last condition is the most important adjustment of the substage condenser and failure to observe it is the most common cause of poor microscopy. If the condenser is only delivering a cone of light of very small aperture, then the objective cannot deliver the resolution of which it is capable (and for which so much money has been expended!). The image in such a case is said to be 'rotten'; although it does have increased contrast, it nevertheless contains little detail and membranes may appear artificially doubled. This is illustrated in Fig. 13.2(A) and (B). Micrograph (A) was taken with the illuminating aperture diaphragm set correctly, whereas (B) had this aperture restricted to a fraction of its correct value. It is thus obvious that correct setting of the condenser diaphragm is of the utmost importance. It may be checked by inspecting the back focal plane of the objective, either by removing the eyepiece and looking down the tube or by using the Bertrand lens, which may be fitted in some magnification-changer

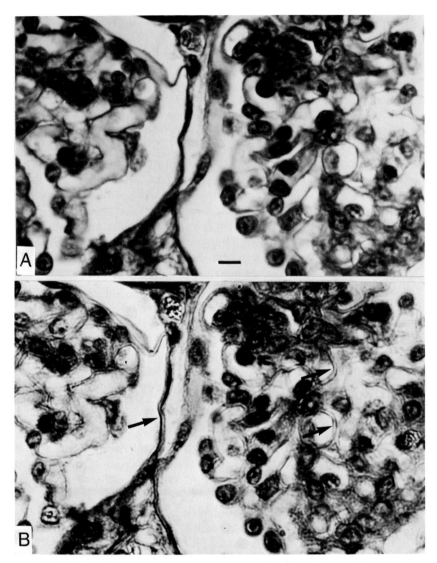

Fig. 13.2. (A) Part of the kidney cortex photographed with an oil-immersion objective used at its correct numerical aperture. (B) The same field photographed with the same objective but now used with the aperture diaphragm of the condenser closed to reduce the working NA of the system. Notice that many of the membranes appear doubled (arrows), so making the structure impossible to interpret despite the increase in image contrast. The bar represents 10 μm.

units to allow the alignment of phase-contrast annuli. On some photomicro-scopes the condenser illuminating aperture diaphragm has a scale engraved in either arbitrary units or directly with the NA of the cone of light that it delivers. In such cases it is only necessary to check the back focal plane once. Recording the correct aperture settings for each individual objective allows them to be reproduced with ease at any future date by reference to the engraved scale.

When incident illumination is in use, the objective acts as its own condenser. The requirement for focusing and centring the condenser obviously no longer applies, but it is still important to set the illuminated field aperture so that glare is controlled and to set the aperture diaphragm to control the NA of the illuminating light. It should be noted that, when using an incident illuminator, the relative positions of the illuminated field and the aperture diaphragms are reversed as compared with a transmitted light instrument fitted with a substage condenser. When transmitted illumination is used, the field diaphragm is found in the illuminating system before the aperture diaphragm (which is located in the front focal plane of the condenser). With an incident illuminator, however, the field diaphragm is located after the aperture diaphragm in the optical train.

Contrast is conventionally obtained by the use of stains or *in situ* hybridization detection systems. In such instances, the microscopy and photography are relatively straightforward. If the tissue is unstained (e.g. *in situ* preparations), then phase-contrast or differential interference contrast microscopy may be used. With the advent of fluorescent labels (for nucleic acid probes), the fluorescence microscope has become more important. With this instrument, as the image is self-luminous and the back-ground is dark, contrast is seldom a problem. Photography of an autoradio-graph (ARG) is often more difficult. In a typical ARG the developed silver grains lie in a plane above that of the specimen. If a high-aperture objective is in use, it is often almost impossible to record both in a single micrograph taken with conventional transmitted light. If the objective is focused on the specimen, the grains are out of focus, and if the grains are focused then the stained background often makes them difficult to recognize. One solution to this has been described by Rogers (1965). It involves taking a micro-graph of the specimen with conventional transmitted light and at the same time using incident dark-ground illumination to show up the developed silver grains. If the relative intensives of the two illuminations are correctly balanced, excellent results are obtained, with the grains appearing as bright specks on the darkly-stained section (see Chapter 3). Standard vertical illuminators will only give a satisfactory incident-light dark-ground effect when used with oil-immersion objectives of high NA, whereas incident-light dark-ground illuminators (of the Leitz 'Ultropak' pattern) work well at low magnifications.

There are now so many different types of photomicrographic equipment in use that it is not possible in a short chapter to consider them in detail. When the microscope is arranged for correct visual observation, the magnified image is virtual and should be located at infinity. In order to record this on film it is therefore necessary to convert this image to a real one located at a finite distance from the exit pupil of the eyepiece where the film may be placed in a suitable darkslide or camera. In theory, it is possible to use a conventional camera mounted directly over the eyepiece, arranged so that the exit pupil of the microscope eyepiece coincides with the nodal point of the camera lens. In practice this is not a very convenient approach. It is also possible to adapt a single lens reflex camera as used for conventional photography for photomicrography by removing the lens and fixing it to a simple microscope adapter tube that carries only the microscope eyepiece. A real image is obtained and focusing is effected by altering the microscope focus, using the camera viewfinder to control the process. This technique suffers from the drawback that it is difficult to be certain of the point of exact focus, especially if any appreciable magnification is in use and that, with high-aperture microscope objectives, focusing in this way will affect the correction of spherical aberration and will markedly degrade the image quality. To counter this, some microscope adapter tubes carry an extra lens that is designed to compensate for the changes introduced by the refocusing necessary to produce a real image. Such systems are only to be recommended for use as a last resort. Much good photomicrography may be done using a simple 35 mm camera attachment fitted to the routine microscope. Excellent photomicrographs may be taken with such instruments. A typical attachment of this type will clamp onto the single eyepiece tube of a trinocular microscope head and will carry some form of leaf shutter and a film-transport mechanism. There is usually a beam-splitter cube fitted below the shutter so that a proportion of the light passes into a side-arm telescope that contains the framing and focusing graticule. The beam-splitter may remain in the beam path at all times so that the image may be observed throughout the exposure. Alternatively, there may be a mechanism that removes the beam-splitter from the light path, so allowing all of the image-forming light to pass to the photographic film. This gives the advantage of much shorter exposure times—a matter of importance when phase-contrast or fluorescence techniques are in use. In many such cameras, the film advance and the determination and setting of the correct exposure are the responsibility of the operator. Some are fortunate in possessing an automatic photomicroscope in which the optical elements needed to form the real image and the film transport mechanism and shutter are all built into the main body of the microscope itself. In such instruments many of the photographic functions (e.g. exposure counting, film wind-on, exposure determination, and shutter operation) are automated and adjustments by the operator are limited to setting the film speed. Whatever equipment is in use, it is essential to obtain a good microscopical image before attempting to photograph it. Some of the different camera types are surveyed in Thomson and Bradbury (1987) and in more detail in Delly (1988) or Loveland (1970). Details of the methods of operating individual makes of equipment should be obtained from the technical manuals provided by the manufacturers.

Choice of film size and type

Often the choice of film size is governed by factors (e.g. availability of equipment) outside the control of the microscopist. If there are no such constraints, then it is worth considering the different film formats that are available in black and white and in colour. The most popular size is undoubtedly 35 mm. This film stock allows an image area of 24 × 36 mm. Almost all photomicrographic attachments provide for the use of this size. Other advantages of 35-mm format are that a vast range of emulsion types is available, both in colour and monochrome, and that the film stock is relatively cheap. Transparencies on 35 mm film are very suitable for projection during lectures. In addition, the processing requirements for this film format are easily available in laboratories and the quantity of solution required is minimal, a factor of some importance if costs have to be watched carefully. Against these advantages must be set the drawbacks that such a small image area needs a considerable degree of enlargement at the printing stage and that consequently extreme care is needed in the darkroom technique, especially with regard to cleanliness.

Roll-film formats have achieved some degree of popularity in recent years as they allow an image area four times greater than that of the 35-mm format. Roll-film cameras are extensively used in commercial photography, but the range of photomicrographic attachments that will accept this size is rather limited at present. In addition, they are inevitably much heavier and of greater bulk. The range of emulsions suitable for use in the technical sphere is not as great as in other formats and very few projectors will accept the 6 × 6 cm 'super slide' transparency format.

If the ultimate in quality is required, then the 5 in. × 4 in. cut-film format should be seriously considered. As the area of the image is approximately 14 times that on a 35-mm film, the degree of subsequent enlargement is much less and grain never causes any problem. A second advantage is that it is easy to shoot and process single exposures. This means that individual treatment may be given to each exposure, so allowing much greater control over contrast, for example. The range of emulsions is large and now many types of photomicrographic equipment may be fitted with a suitable cut-film holder for this format. Marked disadvantages, however, are that the photographic material is much more expensive and a darkroom is almost essential if the microscopist is to load and process sheet film. Although this format produces excellent quality transparencies that are suitable for the production of colour plates, it does not easily lend itself to the production of transparencies for projection.

When the format has been decided, the type of emulsion to use must be considered.

Monochrome

Most of the micrographs intended for publication are taken using monochrome. Very many emulsions are available and the choice of the microscopist is often irrational and governed by, for example, the previous practice of the laboratory. Factors that must be considered in making a rational choice include speed (which is often linked to the contrast and exposure latitude), grain size, and spectral response.

The speed of a film emulsion reflects the amount of light required to affect the silver halide crystals and produce a negative of suitable density. If the quoted speed of one emulsion is higher than that of another then it will require a shorter exposure (the product of the *illumination* and the *time* for which the emulsion is exposed to the light) to produce a negative of similar density. There are several factors affecting the speed of an emulsion, including the intensity of the exposing light, its colour, and the type and duration of the development process. The speeds quoted by the manufacturers are only intended as a guide and it is essential in photomicrography, even with automatic cameras, to make a calibration test to establish the value setting for the other parameters of the system. Unless the level of illumination in the specimen plane is very low (as might be the case in high-power phase-contrast microscopy, or when weak fluorescence is to be recorded) there is no advantage in photomicrography in using a fast film. This is because in many 'fast' films (i.e. with an ASA speed rating of 400 or above), the developed grain size tends to be large and the grains clump in development (compare A and B in Fig. 13.3). Such fast films also tend to give negatives in which the contrast range is low; many of the 'slow' films (i.e. with ASA speeds rated at 32 and lower), on the other hand, tend to produce negatives with excessive contrast when used for photomicrography. These latter films also suffer from the disadvantage that their exposure latitude is very low and this makes it difficult (even with the advantages of automatic exposure) to produce negatives of consistent density.

Recent advances in film technology have resulted in monochrome films that now show virtually no grain clumping and may be used to produce negatives of almost any desired degree of contrast. The use of one of these films is strongly recommended for photomicrography. It is important that the potentialities of one emulsion are learned fully, so that it may be used to the best advantage. The standardization of technique, coupled with keeping of comprehensive records, is essential for successful photomicrography.

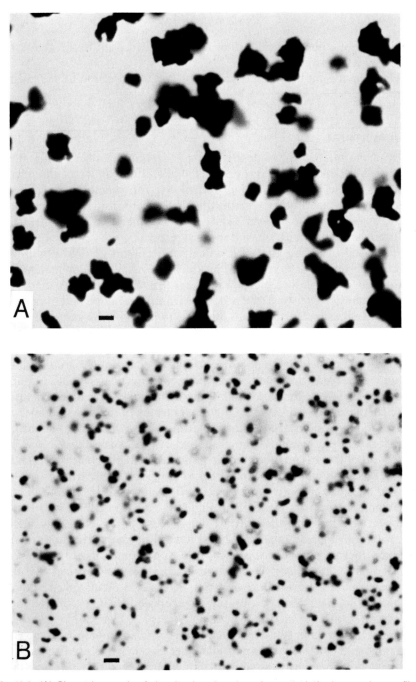

Fig. 13.3. (A) Photomicrograph of the developed grains of a typical 'fast' monochrome film. Notice their size and tendency to aggregate. (B) The developed grains of one of the specialized monochrome emulsions (Kodak Technical Pan 2415) at the same magnification. The grains are much smaller and do not clump, thus leading to a print that is almost without detectable graininess. The bar represents 1 μm.

Colour

If colour is required, the choice involves first deciding whether transparencies or prints are needed. In general photography, when prints are wanted it is often the case that a colour negative film is used. This renders the colours on the negative in a complementary form. Control over the colours takes place during printing, which is often done by automated processors. This gives very satisfactory colour rendering for most amateur photographs, but in the case of much scientific work the machine operator has no reference basis and strange colour casts are often produced. For this reason it seems preferable to shoot the original images onto transparency or reversal film. This gives an image that is (if the microscopy and photography are correct) a true representation of those in the specimen. If colour prints are subsequently required then these may be obtained from the reversal transparencies by a process such as Cibachrome. In this case the printer may use the original transparency as a reference to ensure correct colour representation in the final print.

Contrast control in monochrome photomicrography

One of the major difficulties in biological photomicrography is obtaining sufficient contrast. There are several different ways in which contrast may be achieved for monochrome photomicrography: optimization of the specimen preparation and/or microscopy, use of a filter to record the image in monochromatic light, use of a high-contrast emulsion, or variations in the exposure/development. It is often easier to enhance contrast at the stage of preparing the specimen by the careful choice of a stain than during the subsequent microscopy or photography. Again, the use of a suitable microscope contrast technique such as phase-contrast or Nomarski differential interference contrast may help in this respect, as will correct setting of the aperture and illuminated field diaphragms of the microscope.

If the microscopy is optimized and if the contrast of the image of a stained specimen is still insufficient, then further improvement may often be obtained by the use of a colour filter in the light path of the microscope. Monochrome illumination will enhance the definition of the image produced by most microscope objectives and if the highest possible resolution is sought then this is recommended. More importantly, however, colour filters may be used to lighten or darken particular colours in stained preparations and it is this which makes their use so valuable to the scientist. Colour filters suitable for this purpose may be a sheet of dyed glass, polycarbonate, or gelatine; alternatively, they may be constructed from a series of metallic layers vacuum-deposited on to a glass carrier. A colour filter transmits light of its own colour and absorbs light of other wavelengths; it follows that if we wish to lighten the reproduction of any colour, a filter of the same colour should be used. Conversely, to increase the contrast by darkening a given colour a filter of the complementary colour is required. The basic properties of colour filters used in this way may be summarized as follows.

Filter colour	Darkens	Lightens
Blue	Red, green	Blue
Green	Blue, red	Green
Yellow	Blue	Yellow
Red	Blue, green	Red

The strength of these effects will depend upon the spectral transmission of the actual filters and, although the transmission curves are published (e.g. in Kodak publication B3, 1981), often only trial and error will allow the effect in any given microscopical situation to be determined.

The use of narrow-band colour filters will affect the exposure and it is likely that these will need considerable increases in order to compensate for the light absorption. Such an increase (the filter factor of the manufacturer) often has to be determined empirically. Even with automatic-exposure cameras the suggested exposures may not be reliable, since the colour sensitivity of the photocells used by these instruments may not be appropriate. A full account of the use of colour filters in microscopy has been published elsewhere (Bradbury 1985).

The other major area in which contrast may be controlled is in the photography. As already indicated, the use of a slow-speed film will usually go hand in hand with a great increase in the contrast. This may be illustrated graphically from the characteristic curve of the negative material. When a film is exposed to light and developed for a given time, a degree of blackening of the film will result. If the density (i.e. the logarithm of the reciprocal of the transmission) of the film is plotted as ordinate against log exposure (i.e. the product of the illumination and the time) on the abscissa, then a curve shaped like curve A of Fig. 13.4 is obtained.

This is the characteristic curve of the film and it enables the effect of any exposure to be predicted for that film and photographic developer and development time. The features of this characteristic curve are covered in detail in text books of photography (Stroebel *et al.* 1986; Jacobsen 1978). One feature of interest in the present context is the long straight-line portion in the middle of the curve. This is the region where density differences correspond proportionally to brightness differences in the image. The slope of this portion of the line is given by the tangent of the angle it makes with the log-exposure axis, and is a measure of the rate at which the density increases as the log exposure increases. This is called the contrast or 'gamma' of the film. In recent years a new criterion (the 'contrast index') has been introduced by the research scientists at Kodak (Publication F-14, 1981). Contrast index is the average gradient of a

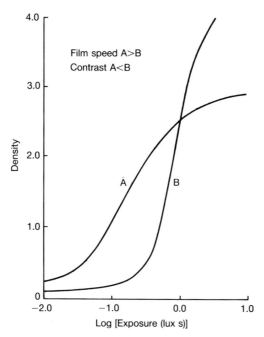

Fig. 13.4. Characteristic curves of two films (A and B) given identical development. The sigmoid shape is similar but there is a great difference in the slope of the straight portions of the curves. Film A would be faster and have much less contrast than film B. It would however, show a much greater latitude in exposure.

straight line drawn between two points on the characteristic curve that represent in practice a low point where arcs having a radius of 0.2 and 2.2 log-density units, respectively, struck from a common centre on the base-plus-fog axis intersect with the characteristic curve. In scientific work, however, where films having long straight-line sections of the curve are in use, gamma is still frequently used to compare the contrast of two films. Curve A in Fig. 13.4 would be for a film of average speed (say 400 ASA) and suitable for general photographic use. It would be capable of recording many different grey tones between the maximum and minimum densities. A film with a curve such as B in Fig. 13.4 would, on the other hand, with the same development routine produce negatives with very few grey tones between the minimum and maximum densities. Such a material would have a much higher gamma or contrast index and be more suited to copying line originals, i.e. those in which only 2 tones (pure black and white) are represented. The film depicted by curve A would have much more latitude in exposure, whereas that of curve B would show little latitude and it would be easy to under- or overexpose using such a film and so produce unusable negatives. It is clear, therefore, that the choice of a suitable film is most important in successful photomicrography. The characteristic curves of Fig. 13.4 show that the contrast achievable with any negative material is largely dependent upon the characteristics of the emulsion itself. It is important, however, to appreciate that a considerable degree of contrast control may also be achieved by altering the type of developer used, the time of development, and to some extent the temperature at which development is carried out. Most manufacturers give characteristic curves for the various developers that may be used with their films, and with some films (e.g. Technical Pan from Kodak; Kodak technical leaflet P255) it is possible to achieve very large variations in contrast in this way (for details, see the manufacturer's technical literature or Thomson and Bradbury (1987) for a brief summary).

Contrast may also be increased by shortening the exposure and increasing the time of the development, but if this course is adopted it must be remembered that there is a tendency for the developed grain size to increase as the development time is increased. If a contrasty, slow-speed film is used for photomicrography, it is unlikely that the final graininess of the negative will be a problem even if large enlargements are produced. It is important to remember that if variations in developer formulation and/or development time are used to control the contrast, then the effective speed of the film may alter dramatically. A film from Kodak (Technical Pan 2415) illustrates this well; if used with the Kodak developer HC110 at dilution B and developed for the specified time and temperature it would have an ASA rating of about 160. If a much lower degree of contrast were required then the developer would be changed to Kodak Technidol LC. In this case the effective speed would be reduced to about 25 ASA. Clearly, with such wide variations possible, it is important to standardize by using one or two film types and keeping the processing schedules as constant as possible. This is one reason why the calibration of automatic photomicrographic cameras is so important.

Given a satisfactory negative, it is still possible to effect a considerable degree of control over the contrast range of the final print by variation in the printing. Printing paper is normally available in several grades of contrast ranging from very soft (grade 0) to extremely hard (grade 5). A normal negative correctly exposed, if printed on a soft paper will give a print with no real blacks but many shades of grey. A print from the same negative on an extremely hard paper would yield an image in which there was black and white but a lack of grey tones so giving what is often called the 'soot and whitewash' effect. Such a correctly processed negative would give the optimum result from printing on a 'normal' paper, which is classed as grade 2. Currently some manufacturers produce a printing paper of variable contrast in which the final contrast of the positive is controlled by varying the colour of the exposing light by means of filters placed in the enlarger light path. This allows great flexibility and is a convenient system because stocks of the extreme grades of contrast paper (which may be required at infrequent intervals) need not be held.

Calibration of equipment

Whatever type of photomicrographic equipment is in use, it must be calibrated in order to determine both

(1) the magnification of the images produced at the film plane by each objective/eyepiece combination;

(2) the correct exposure for the film

Magnification calibration

This need only be performed once as it remains constant unless the optics are changed. In practice, a high-quality stage micrometer is photographed onto high-contrast film using each objective/eyepiece combination which is available. When the film is developed and fixed it is then filed for future reference. When prints are subsequently made from negatives taken with a given objective and eyepiece, a further print is made from the relevant frame of this magnification strip *before* the setting of the photographic enlarger is changed. This will then allow a measurement bar representing a relevant length (usually either 1 or 10 μm) to be drawn on to the micrograph of the specimen. Such a method of indicating magnification automatically includes the magnification increase introduced by the photographic enlargement and is preferable to stating the magnification in the caption of a published figure. With this latter method the reader is never sure whether the illustration has been reduced during the printing and if so, whether the relevant adjustment has been made to the stated magnification.

Exposure calibration

The calibration for correct exposure is important even if a fully automatic photomicroscope is available, since the individual working conditions (i.e. type of film, type of specimen, developer type, and development duration) may render the manufacturer's calibration settings inappropriate. The procedure for preparing an exposure calibration strip differs according to the equipment. If a fully automatic photomicroscope is in use, then a simple check will suffice. A length of the film to be used is loaded into the camera and a well-stained section or an example of a typical specimen is focused. A series of exposures is then made, one at each value of the ASA speed-setting control. The strip is developed under the conditions that are to be the standard, fixed, washed, and dried. The strip of negatives or colour transparencies is then carefully examined on a light box and the setting of the ASA speed that produces an image of appropriate density is taken as the standard for that film type.

If a simple eyepiece camera attachment is in use, then the calibration procedure depends upon the method adopted for giving a correct exposure. In all cases a photometer probe must be inserted at a standard location in the microscope system (often replacing one of the viewing eyepieces in the binocular head). When the microscope is correctly adjusted, the meter attached to the photometer will indicate some arbitrary reading. For black and white photography the intensity of the lamp may then if necessary be altered to make this figure some convenient value. A series of exposures spanning the range allowed by the shutter is made, and the film is processed. On examination it is then a simple matter to determine the correct exposure for that particular level of illumination, film, and developer combination. For future photographs the level of illumination is reset to the same level and the exposure determined from the test strip should give an acceptable result. Bear in mind that this method is only valid if the light level is always measured *at the same point in the optical system*. This method of measurement is of the illumination over the whole field (integrated field measurement) and gives reliable results in most cases with stained sections that occupy the majority of the field of view. It is less valuable if the specimen occupies only a small area in the centre of the field or if fluorescence or dark-ground illumination is in use. In these instances the measurement of the illumination over a small central area of the field ('spot' measurement) generally gives more reliable results but requires special equipment. In some of the more advanced systems, spot or integrated field measurements are available at the operator's discretion.

If colour transparencies are required, then exposure calibration should be carried out with the lamp set at its maximum voltage. This is necessary in order to ensure the light is of an acceptable colour temperature. It is then likely that the image will be too bright for comfortable visual observation. The remedy is to insert suitable neutral density filters in the system at any convenient location: reducing the intensity of the light will result in the colour temperature of the light being too low for the correct balance of the film and the resultant transparencies will have a reddish or yellowish cast. If such an unbalance persists even after using the lamp at its maximum setting, then further tests should be performed with colour-conversion filters (such as the Kodak 80 A, B, C or D blue series) in the light path. If the colour transparencies consistently appear to be too bluish, then the colour temperature of the light is too high and it may be reduced most easily and consistently by the use of one or more of the Kodak 85 series of amber conversion filters.

References

Bradbury, S. (1985). Filters in microscopy. *Proc. Roy. Microsc. Soc.*, **20**(2), 83–90.

Bradbury, S. (1989). *An introduction to the optical microscope*, Royal Microscopical Society Handbooks No 1. Revised Edition. RMS/OUP.

Delly, J. G. (1988). *Photography through the microscope* (9th edn). Monograph P-2. Eastman Kodak Co.

Hartley, W. G. (1979). *Hartley's microscopy*. Senecio Publishing Co, Charlbury, England.

Jacobson, R. E. (1978). *The manual of photography* (7th edn). Focal Press, London and Boston.

Kodak Publication (1981). *Contrast index: A criterion for development.* Publication F-14. Eastman Kodak Co.

Kodak Publication (1981). *Kodak filters for scientific and technical uses.* Publication B-3 (3rd edn). Eastman Kodak Co.

Kodak Publication (1985). *Kodak technical pan film 2415.* Publication P-255. Eastman Kodak Co.

Loveland, R. P. (1970). *Photomicrography: A comprehensive treatise,* Vols I and II. Wiley, New York.

Rogers, A. W. (1965). *The microscopy of autoradiographs.* Scientific and Technical Information. E. Leitz (English edn), Vol. 1, No. 1, pp. 62–6.

Spencer, M. (1982). *Fundamentals of light microscopy.* Cambridge University Press, Cambridge.

Stroebel, L., Compton, J., Current, I., and Zakia, R. (1986). *Photographic materials and processes.* Focal Press, Boston and London.

Thomson, D. J. and Bradbury, S. (1987). *An introduction to photomicrography.* Royal Microscopical Society Handbooks No 13. RMS/OUP.

Appendix 13.1

Processing of monochrome film

Monochrome film is best purchased in bulk (in 30 m lengths) and loaded in a darkroom into a so-called 'daylight' or bulk loader. These are available from photographic retailers and store the film in a light-tight container that has provision for the insertion of a 35-mm cassette. After closing the loader door, the requisite amount of film may then be wound into the cassette. This latter operation does not need a darkroom and it allows short lengths of film to be loaded at any time. As often only a few exposures are needed, the loading of a short length from a bulk loader avoids the waste that occurs if standard commercial 36-exposure cassettes of film are used.

After the film has been exposed in the photomicroscope, it should be rewound into the cassette, which may then be loaded in a darkroom into the plastic spiral of a small developing tank. Developing tanks are available very cheaply from any photographic store and once loaded allow films to be processed without further access to a dark room. Loading is now very straightforward, as most spirals in tanks designed for amateur use have some form of ball or ratchet mechanism to guide the film into the grooves. If difficulties are experienced with loading the spiral, this is often due to (a) damage to the spiral, (b) failure to round-off the leading corners of the film, or (c) the grooves of the spiral retaining water from previous use. A little practice with a length of fogged film is recommended before trying to load an exposed film for those who are attempting film processing for the first time. Most small tanks require approximately 300 ml of developer and fixer for processing a single 36-exposure 35-mm film. Having chosen a film (and a processing routine) it is recommended to explore its possibilities fully and not change constantly. Two suitable films are (from Kodak) Technical Pan 2415 or TMAX 100, but all manufacturers provide suitable film stock. Whatever film stock is used, it is essential to calibrate the equipment as described on p. 233. A suitable developer for Technical Pan is Kodak HC110 used at dilution B. This developer is available in two forms, one a very viscous concentrate, the other (more suitable for general laboratory use) a less-concentrated and less-viscous form supplied in 500-ml containers. This latter form is diluted with water in the proportions of one part of developer concentrate to nine parts of water to make the working solution. This is used once and then discarded, so avoiding problems with developer exhaustion or oxidation due to storage. Although individual protocols must be determined by experiment, the following may be used as a start.

1. Expose a test strip of Technical Pan 2415 on a well-stained section at ASA ratings of 25, 50, 80, 100, 125, 160, 200, and 250.

2. Develop in HC110 (dilution B) 6 min at 20 °C. Agitate the tank by inversion for the first 30 s and then give three inversions on every minute.

3. At the end of development, pour developer away and fill tank with either water or a proprietary stop bath for 30 s.

4. Fix in a proprietary rapid fixer, diluted according to maker's recommendations (usually 1 part of concentrate to 4 parts water) for 3 min.

5. Return fixer to its container for subsequent re-use. Remove tank lid and allow running water into tank for about 20 min.

6. Add small amount (*c.* 10 drops) of photographic wetting agent to the tank full of water. Agitate the spiral in this once or twice. Remove film from spiral and hang in a dust-free place to dry. The excess water may be gently removed from the surface of the film with a rubber squeegee.

7. When dry, examine film on a light box and choose the setting of the ASA speed control that gives a negative of suitable density for printing and record for future use.

Index